CAPTAIN'S DISCRETION

Taking Command of Your Life
in the Confusion of a Balanced World

BY
PETER DENUCCI

Kathie,

Thank you for opening doors!

Peter

Captain's Discretion

Copyright ©1996 by Peter DeNucci

Published by
Apollo Publishing, Marietta GA
Contact Peter DeNucci via the Internet:
capndisc@mindspring.com

Produced by
Angel Works, Atlanta GA

ISBN 1-889013-00-5 Printed in United State of America.

First Printing 1996

10   9   8   7   6   5   4   3   2   1

This book is designed to provide information. It is sold with the understanding that the author is sharing personal experiences with regard to the material presented. If assistance is required, the publisher emphasizes the importance of consulting a professional, so as to provide personal attention in needed areas. This book should be used only as a general guide and not as the ultimate source of information on the subject.

The publisher shall have neither liability nor responsibility to any person or entity with respect to loss, for the use or misuse of this book. If you do not wish to be bound by the above, you may return this book to the publisher for full refund.

The aviation stories in this book are written with the intention of presenting general human factor interactions. Each story is the compilation of: several related, deidentified incidents which occurred at airlines that are no longer flying, and is not presented as an actual recreation based on a cockpit voice recording. For reasons of privacy, the names of the people in this book are fictitious.

*For all those
who do not want to wait
for their next life
to start living*

"The Universe is change;
our life is what our thoughts make it."

-Marcus Aurelius Antoninus (180 A.D.)

ACKNOWLEDGEMENTS

There are few experiences more enjoyable than the sharing of synergy between people united for a common cause. I was privileged to witness such synergy during the development of Captain's Discretion. This joint vision was much stronger than simply the desire of editors and publishers to keep a pilot's disheveled writing from being imprisoned in original manuscripts – it was the pouring forth of their true selves, and for this I thank them:

Dawn Barrs – project guidance, friendship
Henry Bloomstein – editing
Carol Aubin – editing
Anita Shuffer – proofing, marketing and seminars
Sunshine McCarthy – seminars, facilitation
Dr. Albert Doss – paradigms on psychology
Renato Longato – Peruvian and Shamanic guidance
Jerry and Ester Hicks, New Frontiers – insight
Dea Martin – hand analogy
Susan Goodall, Ron Towery, Angel Works –
layout design and production
Peter Gebhardt – final cover art
Michelle Gallagher – original drafts of cover art
Gary Semonian, Marscha Cavaliere, and Nicky Elkourie –
photography

CONTENTS

INTRODUCTION

We Don't Know,
What We Don't Know

Since commercial aviation's beginnings, we've regarded the captain of an airplane as some type of "superman" who could not make a mistake – if he did, catastrophe would result. A model of behavioral perfection, strict procedures helped guide him through any treacherous scenario. These procedures were enhanced by unwavering discipline. This served us well in the 1960s and 1970s as passenger travel entered the jet age. Within time, however, we began to formulate data which suggested this way of thinking, or paradigm, might need adjustment. We found we could no longer attribute the cause of an airline accident to "pilot error," we had to look on the captain as *human*. The cause of pilot-performance related accidents soon became known as "human error," which meant we could deal with them in a thoroughly new fashion. We could find root causes, and implement new training and requirements. There was a down side to this revolutionary thinking; many pilots interpreted it as an invitation to chaos. Understandably so! We were all the products of a perfect, though outdated, training program. For some it was hard – if not impossible – to change.

Despite tremendous resistance, a new science was born: Aviation Human Factors, a vast body of information designed to help pilots function in automated cockpits, establish communicative environments, build teams, and enhance aircraft safety. During my eighteen years of flying experience I have been extensively involved in education, flight training, and Crew Resource Management, as related to Aviation Human Factors.

Captain's Discretion utilizes several Aviation Human Factor terms in order to create a new, exciting method for learning. It will take you on a flight, so to speak, similar to the one I took in my personal life. While studying the interpersonal relationships of flight crews, and working with Human Factor experts from around the world, I discovered many similarities in the way people behaved under pressure, inside and outside the cockpit. Not long before, I went through a personal trauma in my own life. I survived a divorce, which dragged on for two painful years. Though it is common today, I still could not believe it was happening to *me*.

Why am I living in my buddy's basement?

How do I explain divorce to a four year old?

Why am I still driving this mini-van around?

Though I attended several counseling sessions, and sought advice from religious ministers, I found myself wanting more than was offered. I began to ask myself many new, astounding questions, and for the first time in my life, I really wanted some answers.

Was I responsible for my pain? Did I bring it on myself?

How could I love a woman so much, only to end up battling with her?

Can we create the life we want, or is it just chance?

Are there forces in the Universe that my upbringing did not teach me?

How do we make love last?

The vision of **Captain's Discretion** is to share with you the results of my personal quest to find these answers. It was a quest which led me to the discovery of order in the Universe and balance in our world. Some of the following pages are inspired by books that have been around for centuries; others are based on the latest Human Factor data from NASA psychological studies. But at the heart of this book are my experiences with people: psychiatrists from Eastern cultures, Shamans from Peru, pharmacists, scientists, psychologists, along with some unusual inter-dimensional spiritual communiqués.

As we enter an age of spiritual awakening, more of us are becoming aware of a Universal Energy prevalent in all life. This in turn is sparking a desire to uncover hidden truths within ourselves, open our minds to new ways of thinking, and thereby expand our horizons.

Part One of this book, *Situational Awareness*, demonstrates that first, we must be aware of our current environment and the forces acting upon us. A proven curriculum currently used in Aviation Human Factors programs, Situational Awareness Training is used to teach pilots how to maintain a clear-eyed perspective in their ever-changing environments. **Captain's Discretion** applies this highly efficient method to everyday life, thereby empowering us with the heightened state of awareness required for understanding the nature of this universal energy flow.

Part Two, *Universal Precepts*, offers easy-to-grasp techniques for focusing this energy into positive aspects of our life, thus enabling us to *attract* desired experiences into our future reality.

Captain's Discretion may read a little differently to you from other books because it is facilitated, which means it is more interactive. In my Aviation Human Factors work, I have found that facilitation is a more effective, enjoyable method for sharing new ideas and experiences. It allows you, the reader, to ask yourself the questions that only you could think of, and thus, find the solutions for which you have been searching.

The Federal Aviation Administration's definition of Captain's

Discretion is as follows-

"*Captain's Discretion: When used in conjunction with altitude assignments, means that Air Traffic Control has offered the captain the option of starting climb or descent whenever he wishes and conducting the climb or descent at any rate he wishes. He may temporarily level off at any intermediate altitude. However, once he has vacated an altitude, he may not return to that altitude.*"

- Airman's Information Manual

When you have completed this book, flip back to this definition of Captain's Discretion. You will probably discover that it has taken on an entirely new meaning, as will the aviation stories that follow.

But before we begin, let's take a look at a few concepts that will be helpful later on.

Situational Awareness – the accurate perception or knowledge a pilot maintains of the operational environment in order to anticipate contingencies and take necessary actions. In other words, having the big picture at all times.

If we are to take command of our lives, we must have an understanding of how our current reality came to be. We must be aware of the energy (or forces) that helped manifest it. With situational awareness, we will have the big picture, and understand our role.

Quite often it is our paradigm, or perception, that actually prevents us from achieving the situational awareness we need. Because we tend to cling to our own model of the world, our reality may not include situational awareness, and therefore, we may never really be able to anticipate what's in store for us. We find ourselves *reacting* to our environment rather than *creating* our experiences with focus and intent.

Ironically, if we lose situational awareness, we never seem to miss it... until it's too late! After all, if you are not aware of something you have, you will not miss it when it's gone. For this reason, pilots are trained to recognize symptoms which may occur with a loss of situational awareness. These are called Red Flags and when they pop up, they warn us that a loss of situational awareness is not far behind.

Red Flags

Fixation: The focus of attention on any one item to the exclusion of all others. Often we fixate on day-to-day events which have nothing to do with where we want to go in life, so we lose the broader perspective.

Ambiguity: Two or more independent sources of information that do not agree. For example, in terms of religious and belief systems, there are so many sources that we can easily get lost in the confusion.

Complacency: A feeling that everything is going well, based only on the fact that we have done it before, in other words, a comfort borne of familiarity. Sometimes, everything is going along just fine, but at other times we may be ignoring an inner restlessness.

Distraction: A sudden turning away from the original focus of attention. We can become so engrossed in putting out daily fires that we forget our life's path.

In effect, these Red Flags signal us that some aspect of our paradigm, or personal reality, needs attention. We can then take the necessary action to fully restore our situational awareness.

I

SITUATIONAL AWARENESS

1

Is Your Paradigm Helping You?

The Boeing 777 is the most modern passenger jet in the world today. It can carry 375 people at speeds up to 588 mph – nearly the speed of sound. It has a cruise range of over 8,300 miles and can fly at altitudes over 40,000 feet, while burning less fuel than any comparable model previously introduced. The engines are capable of developing 100,000 lbs. of thrust each, the highest in the industry. Due to its high level of cockpit automation, the Boeing 777 requires only two pilots.

Today, one of these pilots is Captain Charles Hansen, and for the first time, he will be allowed to actually fly from one city to another. Having just passed his rigorous "checkride," in which he demonstrated his sharp piloting skills, he is now looking forward to a routine flight. With many years of experience flying B-727s, Captain Hansen enthusiastically waits to command all of this new technology.

After getting tower clearance for take-off, Hansen advances the throttles to the stand-up position, and allows several seconds for the goliath engines to spool up. He engages the auto-throttles, thus commanding the computer to fine-tune the power setting. The huge jet responds to the 200,000 lbs. of thrust, and begins accelerating down the runway. In a matter of seconds, Hansen removes his hand from the nose wheel tiller, and begins steering with his feet. The rudder responds, allowing him to keep this speeding 200 tons centered on the runway. In fact, he can even feel the recessed centerline lights making that familiar thump against the nose wheels.

When the airspeed indicator reaches 137 knots, First Officer Steve Willis calls out, "V1." Hansen now knows he's reached what is known as the "decision speed," which means there is not enough runway to abort the takeoff. From this point on he is committed to getting the B-777 airborne.

Following procedure, Willis calls out "Rotate." This is Hansen's signal to begin easing back on the yoke to lift the big jet off the runway.

As they climb through 1,000 feet, Hansen calls out various commands to his first officer. Following orders, Willis retracts the plane's flaps and updates the Flight Management Computers. They also check with Air Traffic Control (ATC) about the required altitudes and headings.

Climbing out of 5,000 feet, Hansen initiates a right turn to a heading of 90 degrees. He accelerates to 250 knots, and levels off at 7,000 feet. This action maneuvers their aircraft into the clouds, and they are now flying solely on instruments.

ATC: "Twin jet 211, fly heading 110, and climb and maintain 9,000 feet. You have traffic at two o'clock and 15 miles at 10,000 feet. Expect clearance to flight level 230 when clear of traffic."

Willis: "Roger, Twinjet 211."

As Hansen begins his climb to 9,000 feet, he suddenly notices the generator fault light for the number two engine is illuminated. This means the aircraft has lost half its normal electrical source. Though the generator on the number one engine will pick up most of the load, he needs a back up source of electrical power.

Hansen: "Steve, start the auxiliary power unit (APU), and call out the emergency checklist procedures."

Willis: "I'm starting the APU now. Standby for the generator failure checklist."

ATC: "Twinjet 211, you have traffic 1,000 feet above you. Do you have them in sight yet?"

There is no answer on the radio.

The flight attendant's call bell breaks the silent concentration in the cockpit.

Hansen: "Ignore that bell for now, just read the checklist, and when you're done, tell the company we're returning to Portland."

Willis: "OK, Generator Failure Checklist. One–APU start and on line. Two–Deselect inoperative generator. Three–Select..."

While Willis continues reading, the flight attendant's call bell rings again.

ATC: "Twinjet 211, traffic two o'clock, four miles, 10,000 feet. They will not be a factor. Go ahead and call Seattle Center on frequency 125.7 and they'll give you a higher altitude as soon as you're clear of that traffic. Have a good flight."

Hansen is not overly concerned with ATC at this moment; he knows the traffic is 1,000 feet above him, and they have a safe, legal separation. He is concerned with the current emergency. His first priority is to keep the airplane flying!

Hansen: "Continue with the checklist..."

Once again, the flight attendant's call bell rings.

Hansen is now starting to feel the pressure. Everyone is calling on him at once. He turns on the auto pilot allowing the plane to fly itself, so he can focus on getting electrical power back up to the normal output.

Willis: "Verify volts on... ah, Hansen, did you verify the volts...?"

More ringing.

Hansen: "Damn it, answer that flight attendant call bell so they stop calling us! We have an emergency up here!"

Willis: "This is Steve, we have an emergency. We are going to return to Portland. Get the passengers ready."

Junior Flight Attendant: (trying to conceal her nervousness) "Steve, the galley power is out, and the lights on the left side of the plane don't work. We'll get everybody ready!"

Hansen: "All right now, let's finish that checklist!"

Meanwhile the flight attendants are meeting in the forward galley to employ their emergency procedures.

Lead Flight Attendant (to the junior flight attendant who called the cockpit): "I know the captain told you we have an emegency, but did he specifically tell you to prepare for an evacuation and what the brace signal would be?"

Junior Flight Attendant: "No, he just said we have an emergency and are returning to Portland. They sounded really busy up there."

Lead Flight Attendant: "Let's start preparing the cabin now. I'll have to talk to the captain eventually, to find out if he wants to order an evacuation when we land."

Willis: "Five—electrical output..."

Willis and Hansen finish the checklist, and the airplane now has sufficient electrical energy to run all the systems normally.

Hansen: "All right, we are back in business. Now, did you call company operations and tell them we are coming back?"

Willis: "No, but I'll do it right now."

The ringing starts again.

Hansen: "I thought you already told the flight attendants what was going on! Tell them again—we have an emergency!"

As Willis selects the intercom to talk with the flight attendant, he hears the sickening computer yelp of the ground proximity warning system.

"TERRAIN! TERRAIN! TERRAIN!"

"PULL UP! PULL UP! PULL UP!"

Though the pilots are initially confused, they both yank back on the yoke, hoping to climb above whatever is out there; after all, they were still at 9,000 feet. What was going on?

As the airplane hit Mount Rainier, the cockpit shook a little, and the altimeter wound back down to Portland airport elevation. The cockpit door opens, and the calming voice of Captain Paul Kennedy fills the void. "Well, Hansen, looks like you just flew a B-777 into Mount Rainier. Let's get out of this simulator and go to the debriefing room."

This airline had installed a video-computer based debriefing station. With the aid of the new technology, Hansen would be able to see exactly what he did, and the consequences of that action. He would actually be able to see the *cause-effect* of Human Factor related cockpit communication.

During this debriefing, a pilot sees first hand what happens when situational awareness is lost. He will experience it. In this case, Hansen lost situational awareness and flew into the mountain. The Red Flags were *fixation* and *distraction*. Hansen will learn to recognize these Red Flags, as well as the importance of utilizing his crew to their full capability.

There will be no one telling him that his crew resource management skills needed help. He will be able to watch himself on the digital recording as he interacted with other crew members. Afterwards, he will come to an understanding that under the high pressure and intense stress of the first flight emergency, he reverted back to his lone-ranger instincts. Hansen tried to handle everything himself, while giving orders that his team could not follow.

It will be a true self-debriefing for Hansen. He will shortly be offered new Human Factor skills—*now that he has decided he wants them.*

The cause-effect relationship of human behavior is an interesting phenomenon. Whenever we do or say something, there is a response from our environment. When a certain behavior elicits an immediate effect, it is easy for us to see this cause-effect relationship.

When the response is not immediate, we sometimes lose the correlation between cause and effect. We find ourselves wondering why certain things are manifesting in a particular fashion.

A good analogy is thrust control in the cockpit of an airliner. When a pilot moves the throttles, it can take up to ten seconds for the engines to develop the selected thrust. The huge size of the turbine engines require this amount of time to spool up. A well trained pilot knows the timespan between throttle movement and engine thrust, i.e., the necessary cause-effect relationship.

This is why pilots are trained to understand the results *of any input,* into their flight. Knowing that certain inputs will take longer to manifest than others, especially those related to human factors, is essential to understanding situational awareness.

Situational awareness is an ally which will allow anyone to implement the corrections on their flight plan and get them to their desired destination. Being alert to the Red Flags will help maintain this state. Remember, we may not recognize the loss of situational awareness, until it's too late.

2

Fixation

It is another serene evening in South Florida. The weather is perfect and the Jumbo L-1011 jet is cleared for final approach into Miami International. A tired crew has just finished safely navigating their way back home. This is the landing all three pilots have been waiting for, as its completion means the end of a four day trip and time with their families. Everyone wants to get home.

Captain: "Miami tower, this is Tri-Jet 301, on the visual approach for runway niner."

Miami Tower: "Tri-Jet 301, this is Miami tower, you are cleared to land on runway niner."

Captain: "Roger, Tri-Jet 301 is cleared to land on runway niner."

The first officer selects the gear handle to the down position. Only two of the three green lights become illuminated. The crew cannot verify that all the wheels are down – they will have to execute a go-around, and circle the airport for another attempted landing.

Captain: "Ah, Miami, it looks like we have a gear problem... probably just a light bulb. We'll have to go around and take a look at this."

Miami Tower: "Roger Tri-Jet 301, fly heading 090... and climb to 2,000 feet."

Captain: "Roger. Climb to 2,000 feet, heading 090. Tri-Jet 301."

The big jet heads out over the Florida Everglades, allowing the crew more time to solve the problem. The captain levels the L-1011 at 2,000 feet and activates the auto pilot. Now they are able to troubleshoot.

As any experienced pilot knows, most landing gear problems are not actually in the gear itself, but in the indication system, such as a switch or a simple light bulb. This well seasoned crew takes the appropriate course of action and attempts to replace the light bulb first. If successful, they will obviously not have to employ more extensive emergency procedures.

But something unexpected happens during this relatively easy procedure. The first officer cannot remove the snap-in bulb! The captain joins in the effort but with no success. Because of the immediate nature and apparent simplicity of the emergency, both pilots and the engineer are soon trying to remedy the situation. All their energy is focused on one small problem. Fixation occurs and situational awareness decays. While working on the light bulb, one of the pilots gently bumps the yoke (flight controls) and starts an unrecognizable descent into the treacherous Everglades below. The result is catastrophic. Tri-Jet 301 leaves Miami's radar screen and is never heard from again.

Fixation can be deadly!

Though this accident resulted from the loss of situational awareness, the initial cause was fixation. The crew was not aware of their airplane's collision course with the everglades because all of their attention was focused on one aspect of their environment. They lost sight of the big picture.

Until you are intimately familiar with what situational awareness feels like, you probably will not recognize its absence. This is why we have warning signs and Red Flags. When a pilot dwells on one procedure for too long, he becomes aware that he is fixating. More importantly he realizes that loss of situational awareness is not far behind. So a pilot is trained to recognize *fixation*, thus allowing him to take the necessary action in a timely fashion.

Fixation, however, is not limited to the cockpits of airliners. It finds its way into our daily lives as well. Whether it manifests itself in our business world, or personal relationships, fixation always leads us down the same road and creates the same result: the loss of situational awareness!

Let's take a look.

* * *

Bob and Cindy have been married for five years. They are bright, intelligent, fun, and appear to be a happy couple. But they both harbor a whole constellation of seething resentments.

Bob: "Oh, we get along OK. I mean, Cindy is a good wife and mother, but I'm getting tired of her constant shopping sprees. She doesn't cook anymore, and she's becoming a real slob around the house. She just doesn't seem to care anymore."

"But you know, she works thirty hours a week, and we need the money so what can I do? I just wish we didn't have to put the kids in preschool. I guess she needs her space."

"We used to be in love. We couldn't wait to see each other. You know, it used to be so nice when we were first married. Oh, I guess that's just the way it is after you've been married this long..."

* * *

Cindy: "Bob's not that bad. He doesn't go drinking with his buddies every night or anything like that. I just wish he wouldn't play golf all weekend."

"We never get any time together."

"All I do is wait on him hand and foot, and pick up his messes. It's just not that much fun anymore, no romance, no flowers, or even going out to a movie and dinner. It doesn't have to be extravagant."

"The kids miss him, too."

"It used to be so different when we were first married. He cared if I was home or not. He used to plan little outings, just the two of us. Some nights we'd stay up late and just talk. God, we had fun – a bottle of wine, a pizza – we didn't even have money. What happened, what the hell happened?"

"Is this what married life is? Is this it? He doesn't even know I exist!"

* * *

What is happening to this couple?

Fixation on certain aspects of their lives is causing them to lose their situational awareness. They cannot see the big picture while focusing on smaller problems. Which is not to say these small problems do not exist; they are as real to Bob and Cindy as the gear light was to the L-1011 flight crew. It is part of their environment. It needs attention, yes, but not to the exclusion of everything around them.

Bob and Cindy feel as though they have lost control of their lives. In fact, they have begun to do just that. Those aspects of the relationship that they want to cultivate are withering away right in front of their eyes – and they haven't a clue as to why.

They are not situationally aware of who they are and how their thoughts direct energy. They have fixated their attention on something they do not want – each other's undesirable qualities. These same qualities will now get larger because that is where they are focusing their energy, whether they know it or not.

If we do not know what situational awareness feels like, we will not miss it when it's gone. But, unfortunately, we will eventually feel the wrath of its absence.

3

Ambiguity

First Officer: "Looks like we have a line of showers between us and Chicago."

Captain: "I read the weather report before we left and it said there might be some precipitation. We should be able to get a good picture of any heavy areas with the airborne radar."

The DC-9 is now level at 24,000 feet, cruising at 500 mph. Things begin happening fast – maybe too fast. The line of thunderstorms that the captain read about before departure had swelled into a solid line of water, hail, and wind shear. Conditions have dramatically changed.

The weather radar installed on the DC-9 is capable of showing precipitation ahead of its path and alerting the crew of expected ride conditions. Normally, an image of the rain shower would be displayed on the pilot's radar screen, but if the precipitation is too heavy, the radar beam will not be able to penetrate. What shows up on the screen is only a thin, harmless band of rain.

Consequently, the area that appears to be the safest on radar, can very well be the most dangerous to fly through. This is known as radar attenuation.

First Officer: "It looks like a heading of 080 should take us through the lightest part of the storm."

Captain: "I agree. The radar is showing a pretty good ride there. It's kind of funny the way that narrow band of weather is right where we need it. This radar is great."

First Officer: "Sure makes it easy to penetrate a line..."

Within two minutes, the DC-9 slams into a solid wall of water; the golf-ball size hail flames out both engines and shatters the outer panes of both pilots' windscreens. In desperation, the captain wrestles with the flight controls until the last possible moment. A cornfield is picked as the emergency landing area – but later proves to be too rough for the DC-9. The results are tragic.

A crash similar to this happened many years ago when aircraft radar was still in its infancy. Common knowledge with regard to its operation was limited at best. Man simply did not know that much about thunderstorms and their corresponding relationship with radar.

During the early years of the jet age ambiguity was indeed a problem. Pilots experienced the influx of new information, or environmental data, which did not always agree with old data. The eventual loss of situational awareness could then lead to degraded safety margins.

* * *

Anyone who is confronted with two or more independent sources of information that do not agree, can experience ambiguity. For example, in our personal lives a loss of situational awareness can be brought on by religious ambiguity. For most people numerous religions are available, each with different belief structures. Some of us may choose the faith we had as a child; others may venture on a new path after extensive soul searching; some will convert because of marriage.

But will we ever clear up our ambiguity?

In the cockpit of an airliner—where a pilot is constantly processing information—ambiguity must be eliminated as quickly as possible. For example: when dealing with dangerous weather, a captain must check many available sources regarding the flight path. These sources may include the National Weather Service, the airline, Air Traffic Control, on-board radar, and even other pilots. Then, based on all this information, an enhanced decision must be made. Remember, every piece of data received up to this point is true and valid in the *reality of the sender*. It is up to the captain to find the connection, or *universal truth*, behind all these reports so that he can have a safe flight.

The only way to truly clear up ambiguity of any type is to identify one source of information as incorrect, then eliminate it; or, to find a universal truth behind the different sources and come to an enhanced decision. With this type of decision, derived from fully processing new information, we may enter a new paradigm—a paradigm in which there is no ambiguity—leaving us in a state of situational awareness.

With regard to religions, when we discover the universal truth within all those available, then and only then have we eliminated the ambiguity of so many different expressions of humanity's faith.

There is a universal truth that creates a connection between all of us—a powerful stream of energy, uniquely expressed by each person. This energy or "life force" has been given many names throughout history. Every language known has made reference to it in some fashion or another. I *will refer to this flow of energy through us as our Inner-Being*.

When we become in tune with our Inner-Being, we will feel the connectedness with all other expressions of this energy, and ambiguity will naturally dissipate.

As the captain of your life, you have the connection to this Universal Energy, through your Inner-Being. It is a natural by-product of having the "sense of self" and "sense of spirit" that allows you to see balance in the world rather than ambiguity.

In later chapters I will discuss how to be in tune with, or blend with our Inner-Being. First, let's look at a few more examples of why we need situational awareness to achieve this special, clarifying state.

4

Complacency

Robbie is fueling his last plane tonight. Today's total was thirty-five aircraft and 1,640 gallons of 100LL (a low lead fuel). In this small commuter operation he can handle up to 150 Cessna-402 aircraft in one week. The work is tough, dirty, and smelly. But if he keeps up a good pace he can usually make schedule. He also fuels an occasional turbo-prop when it pulls into the gate. Robbie is good at what he does, and the pilots can depend on him to be there with the fuel needed.

In a few moments, Captain Thomas Erickson will need eighty gallons of 100LL – but that is not what he'll get.

As Tom taxies his Cessna-402 into the gate, his passengers are anxious to deplane. As soon as all ten are escorted into the terminal, five more excited travelers are ready for departure. The courteous, busy gate agent begins collecting the new tickets.

"Forty gallons in each tank!" commands Captain Erickson, making sure he is heard above the ramp noise. In a small operation like this, it is up to the pilot to tell the fueler personally the number of gallons he needs.

"OK Tom, you're number three tonight. I'll get you in a few minutes!" Robbie is moving fast. Two turbo-props just came in and needed 100 gallons of JET-A. He had to finish them first. No problem. After all, he'd done this a hundred times.

Tom quickly heads into the flight operations room and takes another look at the weather – just to be sure there aren't new thunderstorms developing. Even though he's been flying all afternoon, it is always a good idea to double-check the latest forecast. The flight release is the next document to review and after verifying everything, he leaves his signature with the operations agent.

Having completed all his preflight checks, Tom straps himself into the pilot's seat and waits for his passengers. The boarding process goes smoothly and before long everyone is settled in, listening to the Cessna-402 safety brief. There are no flight attendants on these smaller planes, so Tom makes all the announcements while taxiing. After receiving takeoff clearance, the little twin engine plane starts down the runway.

With flaps and gear up, Erickson initiates a turn for his first navigation fix. Before he and his passengers even have time to admire the beautiful Caribbean scenery, Tom notices the left engine overheating. As he starts to reduce power, he notices the right engine beginning to overheat. In less than a minute the two engines are so hot that fire is imminent. Tom commands the aircraft to bank left and locks on a heading for an emergency landing back at the departure airport. But, as he attempts radio contact with the tower, the left engine burst into flames. He cannot wait any longer – he has to shut it down, now!

"THROTTLE – IDLE."

"MIXTURE – CUT OFF."

"PROPELLER – FEATHER."

Erickson goes through the drill as he's been trained to do. Except this time, he has to shut down both engines – within seconds the right engine will be so hot the alloy block will start melting! He repeats the drill:

"THROTTLE – IDLE."

"MIXTURE – CUTOFF."

"PROPELLER – FEATHER."

There is no way he can make the airport now; he has to pick a nearby field. Anything about a thousand feet long, no power lines, not too rough...

With just a slight heading change he finds himself set up for a small cow pasture. He will aim for that.

After a successful landing, the right wing of the Cessna-402 cracks open, spilling fuel on the hot engine. The ensuing fire blocks off the right emergency exit, leaving only the two left exits. Miraculously, Captain Erickson is able to evacuate all but one passenger. As he guides everyone to a safe meeting place amidst the blazing aluminum inferno, confusion begins to set in. What happened ?

The answer was easy to find.

While Tom was signing his paperwork in operations, Robbie accidentally fueled his airplane with JET-A – fuel very similar to kerosene, which burns at too high a temperature for reciprocating engines (piston). Both engines literally melted.

Complacency has affected all of us at one time or another. It is a natural human tendency to relax focus and concentration on job tasks, once we have mastered them. If we can recognize this behavior, we can prevent loss of situational awareness. In aviation Human Factors, we study these accidents to *learn* from them. Afterwards, we can install the proper safeguards to prevent their reoccurrence. We do not *blame*; we look for root causes. By studying one accident, we can prevent ten in the future.

* * *

Complacency can creep into our life slowly, concealing itself behind things we do not suspect: creature comforts, relationships, careers, just to name a few. It will keep us at a level of existence where we are not uncomfortable enough to make any changes, yet not comfortable enough to be happy.

Put another way, our current lifestyle can be just good enough to keep us in it. We don't really seem motivated to make any changes. Because it is not all that bad, we tend to leave it in the status quo. This particular type of complacency, if not recognized early on, will result in the loss of situational awareness.

Here is one couple's true story:

Ted and Marcia have been married for eighteen years. When they got married, Marcia discontinued her college education in teaching. Though she had some reservations at the time, she thought it would be best for everyone involved – including any future children – not to work outside of the home. Ted also thought it was the right thing to do, and besides, he enjoyed the role of the young, ambitious breadwinner for his new family.

Within a few years, two beautiful children were born, eighteen months apart. Marcia found her days occupied with child rearing, and Ted was putting in extra hours to make the payments on their new house. They were starting to enjoy the good life.

As the children got older, both parents wrestled with the difficulties of raising them, doing whatever they could to be good providers. By now Marcia was a full-time homemaker, and had little time and energy left at the end of the day. Even if she had the time for herself, Ted would probably be working – he wasn't home much anymore. On several occasions, however, he did suggest moving to a less expensive neighborhood so he could work less, but they could not find one they liked.

Like many couples, Ted and Marcia began spending less and less time together. They lived under the same roof, but could never seem to make that extra time just for themselves.

Sadly, they found that many of their interests had gone in different directions over the years. They were not doing much more than cohabitating.

After the children grew up and left the house, Marcia started to feel somewhat unneeded and unimportant. She regretted that she hadn't been able to get her teaching degree. She felt profoundly unfulfilled. She wanted a separation.

Ted was in shock. He knew that he'd been gone a lot, but also knew Marcia enjoyed the lifestyle he had worked long hours to provide. *In his view*, she had spent the last several years doing her own thing anyway.

Remember, to have situational awareness, we must have an accurate perception of our operational environment, in order to anticipate contingencies and take necessary action. Ted and Marcia let complacency creep in, little by little, year after year.

In their relationship, situational awareness deteriorated slowly over time; therefore, it was unrecognizable. It not only vanished in their personal lives, but also in the marriage itself. They *both* became affected by it.

Ted thought that as long as he kept working, or providing, everything would be fine, and the marriage would be able to continue. After all, he'd been a good husband and a good father. Likewise, Marcia believed that as long as she was a good mother, and ran a good home, the future would take care of itself.

When we recognize complacency in our lives, we must take action and adjust our thinking. We cannot wait until negative emotions begin to motivate us.

When you are the captain of your life, you are tuned into the best guidance system you can have: the human emotional system. Your emotions are linked directly to your Inner-Being, which will signal you either positively or negatively – depending on your thoughts at any given moment. But, without situational awareness, and an understanding of how emotions relate to your thoughts, these signals will not be able to guide you. In fact, they may add confusion to your life.

Because an understanding of situational awareness is a prerequisite of comprehending our emotional system, fully described in part two of this book, let's take a look at another Red Flag that can show up in our lives.

5

Distraction

"The Springfield weather, two thousand overcast, visibility two miles, winds are calm, altimeter 29.97. Runway 9 left approach in use. Taxi way delta and taxi way alpha are closed."

Travis did not have to listen to the remainder of the Springfield advisory tape. He knew those taxi ways were closed; after all, most of his 256 hours of flight time were logged right here in Springfield. He even knew the airport frequencies by heart. Having just passed his Instrument Rating Checkride, he was now legally qualified to fly an instrument approach without an instructor on board. He wanted to log every approach he could get his hands on. Travis knew each minute of actual instrument flight time was pure gold in his logbook. He loved to fly the Mooney airplane on instruments—it was a quick little plane that offered a challenge, a sports car with wings.

Travis: "Springfield approach, this is Mooney seven niner gulf, I am ten miles north at four thousand five hundred feet squawking one two zero zero."

Approach Control: "Roger, Mooney seven niner gulf, squawk code three two four five."

Travis puts the new code into his transponder. This identifies his airplane on the radar screen. Once air traffic control has a positive radar return, he is told to descend to 1600 feet and fly an intercept heading for the approach. Quickly, he reviews the instrument approach plate. He wants to be well prepared for this landing. Those instruments are going to be right on center! As he nears the runway, approach control hands him off to the tower.

Travis: "Springfield Tower, this is Mooney seven niner gulf, on approach, runway 9 left."

No response from the tower.

Travis: "Ah tower. This is Mooney seven niner gulf, runway 9 left."

Tower: "Roger, Mooney seven niner gulf, continue. Expect clearance to land shortly."

Travis: "Roger, seven niner gulf."

As Travis descends through 1200 feet, he struggles to get the airplane back on center. The four seater is doing a pretty good job of staying on course, but Travis wants it nailed; he wants it to be exact. That extra radio call on the approach didn't help things because he had to keep picking up the microphone.

"OK, that looks better. Got her nailed! Now for the landing checklist. Mixture... prop...."

Tower: "Mooney seven niner gulf, I may have to break you off the approach, ah, are you getting a good signal for the approach?"

Travis: "Springfield, this is seven niner gulf, I just lost the instrument landing signal, but I have the runway in sight."

Tower: "OK, seven niner gulf, you are cleared to land on runway 9 right, I say again, 9 right."

Travis: "Roger, cleared to land runway 9 right."

Travis has his hands full now—he had to break off the approach for 9 left and line up his plane on 9 right. He had practiced this sidestep maneuver before, but not alone and certainly not in these conditions. He commands a hard bank to the right, followed by a hard bank back to the left. The new runway begins filling the windscreen. Travis now has to employ the short field landing technique, which means flying at a slower speed and stabilizing the glide path just above the tree line at the approach end of this short runway. The landing point is selected, the speed is locked on 85 knots. His glide path is perfect. As he enters his flair, he slowly retards the throttle, holding the nose off.

Tower: "Hey, Ah, Mooney seven niner gulf, check your landing gear...!"

SCREEEECH!!!

Although Travis nailed the instrument approach and successfully maneuvered the airplane to runway 9 right, he forgot a very important requirement to facilitate a smooth safe landing—he had not extended his landing gear! Distracted several times during the final approach, he never actually completed his final landing checklist. The items that distracted him were *valid and important events*, which are legitimate aspects of a pilot's environment. They required and demanded his attention.

Similarly, distraction can cause a loss of situational awareness for a group of people, whether they are a family, a corporate team, or even a three-man flight crew.

* * *

Stanberry is one of those small towns in the Carolinas that used its Southern charm in the early 1980s to attract new business: boundless land for development, a low tax base, and a country smile to boot. Top it all off with a moderate climate, and you've baited the hook for some new neighbors. And they came running.

The United States military also had plans for Stanberry. President Reagan had decided to increase the military budget, and some smaller army bases needed to be expanded. For Stanberry, this meant a boost to the local economy, and the necessity for increased airline service. Their small local airport would have to be renovated—as would the military airport on the other side of town, at least twelve miles away. But then again, distance is relative to speed, which in turn is relative to time, which is relative to...

* * *

The Boeing 727 streaks across the Carolina night sky, the mountains below offering no sign of a resting place. In the distance, maybe fifty miles or so, there is the faint glimmer of street lights. Traveling at eight miles a minute, the crew would be on top of Stanberry in no time.

First Officer: "I sure wish the Stanberry VOR (ground navigation instrument) was in service, it would make it a lot easier to find this place. I'm going to try to pick up another signal."

Captain: "OK, you try to lock on that and I'll ask approach for vectors. You'd think we could see it by now, being a clear night and all."

First Officer: "This signal isn't strong enough for navigation either; must be the mountains."

Captain: "Approach, we can't seem to get a good navigation signal tonight. How about a heading for Stanberry?"

Approach: "Roger, fly heading one five zero, and you should pick up the airport at twelve o'clock... twenty-four miles."

Captain: "Heading one five zero, and we're looking for the field."

First Officer: "Hey, I think I just saw the airport beacon... there... about twelve o'clock and twenty miles."

Captain: "I have the airport in sight and the runway lights also.

Yeh, north-south configuration, that's it. That's got to be runway one eight..."

"Ding, ding, ding," (the flight attendant's call bell)

Second Officer: "This is Rick, what's the problem?"

Flight Attendant: "There's a weird noise coming from this back service door. Kind of a whistling sound."

Second Officer: "Probably just air leaking through the door seal because we're descending. These old 727s do that sometimes, but we'll check it on the ground anyway... which will be in about five minutes."

Captain:"What did the flight attendant want? Is there a problem?"

Second Officer: "No problem, just a service door leak. I told them we'd take a look at it on the ground—unless you want me to go back now?"

Captain: "No, we can check that on the ground. And besides, we need you up here to help look for some Cessna between us and the airport. Approach is not talking to him and he might be lost."

First Officer: "I got him... two o'clock... I don't think he'll be a factor."

Captain: "Good, tell approach we have the traffic, and the airport in sight. And Rick, you tell operations in Stanberry to have maintenance standing by to have a look at that aft service door. Give me flaps, five degrees."

First Officer: (While selecting flaps five degrees for the captain) "Stanberry approach, we have the traffic and the airport in sight."

Approach: "Roger, you are released to Unicom frequency 122.8. Call me on the ground to cancel flight plan."

Note: Because some airports are not large enough to have FAA control towers, a Unicom or Universal Communication frequency would be used. This is the frequency that pilots use to talk to each other and to notify each other of their intentions.

Captain: "Gear down... landing checklist..."

The pilots complete all pre-landing checks, and the captain rolls the big jet onto the runway for a smooth landing. As he clears the runway and the first officer finishes the after landing checklist, he gets a sickening feeling inside. He stops the 727 on the taxiway and sets the parking brake. Just then, the first officer looks up and shouts, "Hey, why are all these military planes parked here?"

A scenario very similar to this happened in the early 1980s. The crew was distracted during their final descent into a small southeast airport and landed at the wrong destination. The items that distracted them were *valid and important events*. The incidents required and demanded their attention.

Valid and important events can also distract us in our personal and business lives. We can work day after day on important projects which have little to do with our life's true path. These projects, though important by appearance, may not get us to our destination. We can invest years in developing a lifestyle that, in the end, we care very little about. We relinquish our situational awareness because we are too busy putting out fires. Once again, we lose sight of the big picture.

The key is to envision your destination now, before you get there, so you will not get lost! It is the best way to prevent *landing at the wrong airport*, only to find out you don't have enough fuel to go on to your real destination.

* * *

Destination Checklist

Before you read any farther, stop a moment and jot down quick answers to these checklist questions. Some brief reflection will help you check your own state of situational awareness. Your answers will be useful later in the book.

What type of life will you have in one year, with regard to the following areas? Answer as though you are just having a casual chat with a friend. In other words, use your normal vocabulary.

Family / Social:

Career

Health

Leisure

Religious / Spiritual

Now, let's look ahead six months:

Family / Social

Career

Health

Leisure

Religious / Spiritual

Describe your life today in these very same areas.

Family / Social:

Career

Health

Leisure

Religious / Spiritual

Immediately after finishing this checklist, write down three or four words which characterize your *current* emotional state, e.g., happy, sad, excited, anxious, afraid. Write down how you *feel right now.*

1._____   3._____

2._____   4._____

Again, this will be useful when we begin discussing the electro-chemical relationship between your thoughts and your emotional system.

In summary, when we pay attention to our Red Flags: *Fixation, Ambiguity, Complacency,* and *Distraction,* we can maintain our situational awareness, and thus be in a good position to take command of our lives. By knowing where we are, at all times, we can start plotting our course and accepting the responsibility of being our own captain.

6

New Paradigm: New Solutions

The aforementioned aviation stories are based on a combination of incidents that happened many years ago, at airlines that no longer exist. Because we have the ability to study each and every one in great detail, we have been able to develop human factors training programs. Though initially introduced as "enhancement" type training, such as Crew Resource Management, these programs were actually part of a much larger national training overhaul: the Advanced Qualification Program, or AQP.

As these Human Factor skills and philosophies were integrated into the airline training environment, they gradually became the way of doing business–similar to integrating a very complex Total Quality Management program into other industries.

This training, which encompasses situational awareness and other Human Factor skills, is now considered the only way to fly a jet. Under the AQP umbrella, all training and operations will reflect this way of doing business.

Human Factor skills have become so integrated into the norm, or way of doing business, that they are part and parcel of the AQP itself. Now, when a pilot goes through an Advanced Qualification Program, he or she is being trained from the beginning in situational awareness and other Human Factor Skills. Ironically, the old method of training is now becoming foreign. As a captain who regularly flies, I am pleased with the progress made in aviation, and the continuing strides made towards safety. By dramatizing past airline accidents, I wanted to show that the men and women in aviation are always ready to learn from the past. As a result, we never stop striving to implement better training and safety measures.

In this same spirit, we are now going to look at some of the results of my personal quest for an understanding of a Universal Energy. We are going to look at some Universal Precepts, or guidelines, as to the nature of the energy that flows through us all, first through our bodies, then into our future reality. We will discuss how we as human beings can focus this energy on positive aspects of our life, and thereby attract desired experiences.

As I stated earlier, these precepts were developed through my independent research, and are not part of any aviation training, or connected with any type of aviation operation. They were designed primarily to address the problems we encounter in our everyday lives.

II

UNIVERSAL PRECEPTS

7

Precept Number One:

Everything exists at its own unique vibrational state.

*"The foregoing generations beheld God and nature
face to face; we through their eyes.
Why should not we also enjoy an original
relation to the Universe?"*

- Ralph Waldo Emerson

Whatever your current perception of the Universe may be, try to expand it to the limits of your imagination. If you are like most of us, you will find this a somewhat difficult task: Our minds have trouble picturing "all that is."

We may start with a picture in our mind's eye of everything we have seen on earth, then expand it to include things we have not seen but have heard about. From here we will usually jump into the solar system with all its planets and various moons. We must not, of course, forget gravity and other balancing forces that keep all these celestial bodies in their prescribed orbits. *All of these creations, we attempt to encapsulate – in a word.*

From here we try to see an ever-expanding Universe, with billions of planets and stars, and our comprehension begins to boggle. The Universe is getting more grand while we are shrinking in comparison. We may even begin to feel somewhat insignificant, and wonder why we are here. What difference could *we* possibly make?

We can make a tremendous difference. All we need to know is how to get in tune with, "all that is." Otherwise, we will never reach our true potential and find out who we truly are.

We are about to find out that our bodies exist at a unique vibrational state. We will also discover that our thoughts *influence* the life-giving energy flowing through us. In addition, we are already aware of the fact, that with this flow of energy into our body, we have life. Without the flow of energy, there is no life.

Just as our human bodies maintain a vibrational state, our environment exists at its own unique vibrational frequency. Everything around us is a manifested form of vibrational energy. We generally do not think of ourselves as living in a vibrational environment, but if we take a closer look we will discover it.

All light exists at a vibrational frequency, which gives us our perception of colors. The vibration of each color affects us in different ways. For example: The color blue tends to have a calming effect, while red may spark an alarming or exciting sensation. Everything we see is perceived through vibration.

All sound resonates at a unique frequency, which gives us our perception of hearing. Some sounds are very soothing, while others can cause us discomfort. Everything we hear, is perceived through vibration.

All electrically transmitted television and radio waves carry themselves on a certain frequency. We cook our food with a microwave frequency. All these frequencies are around us; we only have to decide which one to tune in to, or block out. (*see chart*)

Some Frequencies (in Hertz) of the Electromagnetic Spectrum

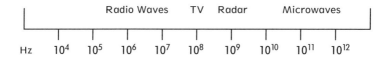

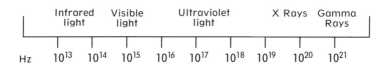

Example: $10^6 = 1,000,000$ Hz = 1Megahertz

The various expressions of this Universal Energy are the stuff of which we and our environment are made. The Universe itself is an unlimited expression of this Energy. We live on a planet that is a unique expression and is in harmony with that flow of energy. When we become in tune with this same flow, through our choice of thoughts, we choose how our expression will be. The choice is ours. If we choose to vibrate at a frequency which is in harmony with this flow, we will be happy.

Our Inner-Being, the connection to this energy flow, is vibrating at that higher frequency on the spiritual plane, and will signal us through our emotions. When our thoughts, at any given moment, are in tune with our Inner Being's extended, enlightened, overall intent, we will receive an increase in the energy flow, and our body will respond with positive emotion, *resulting in happiness*. If our thoughts, however, are not in tune with this overall intent, we will have a decrease in energy flow, and be signaled with negative emotion. The negative emotion is not a punishment for thinking a certain way. On the contrary, it is like a warning from your best friend, telling you what you are co-creating, or attracting into your future, will make you unhappy.

Our Inner-Being knows that our repeated thought patterns in the physical plane (on earth), will eventually attract that future reality. This is why we have this wonderful inner guidance system. It will never fail. Our Inner-Being is aware of the experiences that will make us happy, and will always signal us through our emotional system.

The desired experiences of your Inner-Being, could be very different from what you are doing now in life. But these new experiences will seem very familiar once you start having them. Because they have been attracted by your true self, you will be answering a call from an inner voice, that you may have heard, but never listened to.

When you wake up in the morning, who do you expect to see in the mirror? Obviously you expect to see yourself, but who are you? Are you physically fit, emotionally strong and balanced? If not, who is responsible for that reflection? The answer is simple, you are. More importantly, you are also the Inner-Being generating a reflection, or expression, in that physical body.

In effect, we are Spiritual entities first, we then manifest as human beings through our intertwining electrical, chemical, and physical makeups.

8

Precept Number Two:

Synchronistic vibrational states will attract to each other.

We are learning that our bodies are a prearranged molecular structure resonating at their own unique vibration. In effect, we are a massive "vibrational vignette." But *why* are varied experiences and people attracted into our lives?

Do you remember the intimate relationship between your physical body and your Inner-Being briefly mentioned in the previous chapter? Keeping those thoughts in mind, we will now discuss why it is so important to take responsibility for this relationship.

Use your imagination and envision your Inner-Being as a powerful magnet capable of attracting anything into your life experience. This powerful magnet can take aim at anyone, any thought (think of a giant universal thought bank), any life experience, and bring it into your day-to-day life. Obviously, the Universe is never short on supplies, so anything can be attracted into your life.

Your Inner-Being will attract those people, or experiences, that share your vibrational state, or who are in, what is called synchronistic harmony with you.

Put another way, our vibrational state, fueled by energy through our Inner-Being *attracts* the actual experiences that happen in our day-to-day life.

The power of the Universe is like a mighty river that can deliver anything to us; we only have to "get with the flow" and be tuned into what we want to attract. This river, or energy flow is transpiring ad infinitum and is happening whether we are aware of it or not.

What is our role or opportunity in all this? Our role is to focus on thoughts and experiences in the physical plane (earth), that we want our spiritual magnet (Inner-Being) to deliver to us. The more intense the focus and vision we possess, the quicker the Universe will deliver it to us. By blending with our Inner-Being we can purposefully co-create our future.

There are several guidelines for this visualization process discussed in Chapter 11, *Manifestation Procedures*.

9

Precept Number Three:

Vibrational interaction creates growth.

*"Whenever one life touches another, we help
or we hinder. There is no escape—
man drags man down, or lifts man up."*

- *Booker T. Washington*

As we co-create with the Universe on the physical plane, we find out what we do not want, create what we do want, and interact with other people doing the same thing. Through all of this confusion, we are actually in a state of continuous growth. Which is one reason we may have chosen to incarnate in the first place. As we become closer and more in tune with our Inner-Being, our experience here can be much more joyful.

It is through interaction with others that we learn how to grow. It is through our individually created realities that we experience this growth. We may even experience growth through another person, whom we allow into our reality.

Everyone is creating on the physical plane through their own Inner-Being. All of our Inner-Beings are connected to the same source. The amount of Universal Energy, flowing through us is a matter of choice.

Each and everyone of us is going through this growth process in our own way. Personal experience can be a very conclusive teacher. If we want to help make a more joyous environment, we have to *individually* become the vibrational vignette that we envision. We must become who *we* truly are if we want to be a positive point of focus in another's life.

We cannot give away anything we do not have!

The example of our life is our best asset in any situation. Think back to your childhood for a moment. Who did you learn the most about life from? Was it someone who preached one value but lived another, or was it someone who may not have said all that much, but lived their values? Only you know the answer.

* * *

The cockpit of a jetliner is one microcosmic analogy of human interaction. Each pilot has an internal and unique vision of the flight he is about to share. The goal of standardization of flight procedures is to channel both pilots' visions into one harmonious synchronistic event. These operating procedures are very precise, leaving little room for deviation. But procedures cannot fly airplanes; otherwise, anyone could hop into a jet, pick up a flight bag full of manuals, and takeoff! If this were true, there would be no accidents, because we could write all the best procedures ahead of time!

The procedures, then, serve as a tool to bring both pilots into the same loop so that their unique differences can enhance safety. It is this appreciation of individuality that keeps at least two pilots in the cockpits of today's modern jets. In other words, if one pilot misses something, the other one will pick it up because of their different perspectives.

The same holds true in life. We need to appreciate the differences in people rather than criticize them. By doing this, we not only learn from their perspective, but we keep our positive energy and emotion coming from our Inner-Being, and keep ourselves in that place of attracting what we want.

"But hey, I don't want to learn anything from this guy. He shares none of my values!"

This happens to all of us sooner or later. Whether we are the one having the thought, or it is someone thinking it about us. The real question we need to ask ourselves is, "Why is this person in my life?" Remember, we attract experiences, and people are a major part of our experiences. What we need to try to do is focus on and elicit the qualities that we do enjoy. If we can't find any, it is usually because we are blinded by our own narrow focal point. In other words, once we focus on the negative qualities, as we perceive them, they become larger! Because that is precisely where we are directing our energy!

Our negative criticism of another person's qualities may hurt them (that is their decision), but it will absolutely effect *our energy flow*, and it will be in a negative fashion. We are aware of this immediately, by the way we feel—bad! So, in effect, we are trying to elicit positive qualities from another for our own benefit as well as others. This may seem to be a very selfish way to go about it, but as hard as we may try, we can only attract our experiences in this life. There is, however, true long-lasting benefit for everyone involved. The other person will probably join in and focus on their own positive qualities because you have brought attention and energy to those desired qualities. This is one way we can enhance society. As you begin to vibrationally align your thoughts with these desired qualities in people you meet today, you will soon be attracting different people into your life experience with an *abundance* of these same desired qualities. You may even end up with a new group of friends. It will be up to you to decide how they fit into your life.

Not only is it detrimental to judge others—we are not *capable* of doing it. We simply do not have the ability to know everything others have attracted into their life experiences to create what we see for them at that moment.

This applies to our own life also. When we judge our past actions and dwell on them, we may end up with a guilt complex. This guilt is a negative emotion, a signal from our Inner-Being telling us that the thoughts we are having now are not in tune with our vision. Stated another way, we feel bad today because we are dwelling on something in the past, and our Inner-Being always wants us to focus on the future.

If you remember, many times we must experience what we do not want before we can identify that which we do want. All of us have gone through this. So it's up to us to learn from our past by recognizing, but not dwelling on, those experiences we did not enjoy, thereby empowering us to go forward with a new individual focus and vision.

10

The Power Of Thought Energy

"Where the telescope ends, the microscope begins.
Which of the two has the grander view?"

-Victor Hugo

In our new paradigm, we look at everyday thoughts as the source of our experiences. In other words, our thought energy has a fundamental effect on our everyday existence. We will find that this is different from what many of us perceive as positive thinking. It is much more.

The power of thought energy is currently being studied by some of our most prominent physicists and medical doctors.

Whether you venture into deep space or into the subatomic world of the human brain, you'll find the same elements and structural relationships. The stuff of which the universe is made up is not only the basic thread of our existence as human beings, but our conduit of energy flow.

As we go deeper and deeper into the human mind, traversing neurons and dendrites, we begin to get an understanding of consciousness itself. After all, what is it exactly that gives us our self-awareness, or for that matter, our situational awareness?

Today, physicists and scientists have come up with a twenty-first century picture of one of the lowest levels of brain function, which happens to match some of the oldest spiritual philosophies on the planet. With their electron microscopes and high-tech scanning equipment, researchers have the capability to look deep into our physical existence. Quantum mechanics, one of the representative sciences, delves into the subatomic world of electrons and their related functions to our physical existence. We find through these quests that *single-state vibration* is at the heart of all this activity.

The housing for these structures lies within each and every cell of the human body. Microtubules, as they are known in the world of Medicine, are thin hollow tubes of protein about ten-millionths of an inch in diameter.

In addition to actually creating the foundation and structure of each living cell, this skeleton-type of microtubule network is responsible for transmitting information throughout the body. The meshlike jungle of microtubules does this via several methods, one is the creation of pathways for chemicals which are sent to various regions within the cell. Another is the conduction of vibrational waves not only across the individual cell, but also to downline cells. It is this transmission of vibrational patterns that is getting the scientific community excited.

They have discovered that there is an electron in one of the microtubule molecules which is directly related to consciousness. In other words, this electron is moving in a prescribed pattern whenever the human mind is providing conscious thoughts.

Furthermore, these microtubules have the ability to transmit *and* receive vibrational signals. In effect, they vibrate, or harmonize, with the neighboring microtubule just as one tuning fork will match the frequency of a nearby fork. However, this cellular process seems to be a bit more sophisticated, and effective. The entire vibrational wave is effectively carried without interference from other cellular activity. *The result is a prescribed, single state vibration transmitted intact, to the area that it was directed, without interference.*

In summary: Based on the electron position (electrical thought energy), a microtubule chooses a vibrational state and directs the behavior of the living cell. The microtubule carries signals, or communicates with, neighboring microtubules, thereby achieving cellular communication. This communication is then translated into chemical responses in the body, which allows us to have emotions. The resulting vibrational vignette is next translated into the fundamental construction patterns of cells themselves, thereby becoming the building blocks of our flesh and bone bodies.

11

Manifestation Procedures

1. Be aware of what you do not want.

2. Identify that which you do want.

3. Visualize yourself already there.

4. Allow your environment to align.

5. Physically take appropriate action.

Be Aware of What You Do Not Want

Whether aware of it or not, we have attracted everything into our life experience. We have vibrationally linked up with that which we want and that which we don't want – sometimes referred to as good and bad. Our past experiences in this life serve as our school yard, from which our future experiences can blossom.

Our "bad" past experiences can lead us to cleaner definitions of what we want down the road. More importantly, our Inner-Being may have attracted these experiences to yield environments which help us fine tune our thoughts, focus our energy more effectively and, enable us to discover who we truly are.

In order for us to maximize our "good" experiences, we must first categorically determine which experiences we wish to continue, and which we do not. Because our lives are actually built out of our daily thoughts, we need to be certain that the majority of our thoughts focus on experiences we desire. If our thoughts tend to be negative, we pull in negative experiences.

How many days have you had that seemed to start off bad and got continually worse? We've all had them. "I'm having a rough day," "I got out of bed on the wrong side this morning," "When it rains it pours!" These are some of the ways we categorize these bleak days. When you've had such a day, do you remember dwelling on the negative events, even though you hoped the day would actually turn around?

When you feel bad, it is a signal for you to monitor your thoughts. The key is to recognize negative thoughts for what they are, then replace them with ones that bring positive emotions. These new thoughts will begin prepaving a future for you through the natural process of attraction.

If you are currently having life experiences you do not want, it could be because of those prepaving thoughts and their associated vibrational states, that have attracted those unwanted life experiences to you. This is why it's so important to have focused thoughts. Nothing extraordinary, mind you, just daily thoughts centered on what you want.

To some this may sound simplistic, but actually it is because of some hidden barriers to situational awareness that we lose the big picture, and continue in a lifestyle that we do not want.

"A capital ship for an ocean trip
was the Walloping Window Blind-
No gale that blew dismayed her crew
or troubled the Captain's mind.
The man at the wheel was taught to feel
contempt for the wildest blow.
And it often appeared, when the weather had cleared
that he'd been in his bunk below."

-Edward Carryl (1886)

Distraction is a hidden barrier. It is a tricky phenomenon because it keeps us from recognizing things we do not want and from identifying things we do want. We can certainly spend an entire day in crisis management, putting out fires and keeping ourselves busy. We then have little time to focus on the experiences we really want, and so the cycle repeats itself. We can get stuck in endless, mediocre co-creating, never realizing our true potential.

Henry David Thoreau said it best:

"The mass of men lead lives of quiet desperation,
what is called resignation is confirmed desperation."

The key is to know your current life has been prepaved by your past thoughts. Even more importantly, your future life is being paved by your current thoughts. Your Inner-Being is attracting people and events in the Universe and bringing them into your neck of the woods. If your daily thoughts are random acts that lack focus, then your life will lack direction, because your Inner-Being will not have positive targets for attraction! Ultimately, our thoughts are the true building blocks of our experiences.

Worry is another state of mind we allow ourselves to get into, because we lack faith in who we are. This negative emotion is another hidden barrier blocking the flow of energy that we need to create our future.

We worry about the future when we feel we do not have control over our experience in it. The reason we worry is because our Inner-Being is signaling us that our thoughts are not in tune with our desired future, i.e., a purposefully co-created future. In other words, *we are thinking about what we do not want because we believe we can't get what we do want!*

Identify That Which You Do Want

Once you know what you don't want in life, you can begin to focus on what you do want. Again, this may sound simplistic, but there are two obstacles here as well. The first is obvious: We do not always know what we want—which can be most troublesome. Many of us have been spending the majority of our time managing what we do not want, leaving little time for having thoughts about what we do want.

As we've already seen, the Universe contains everything, and it can be delivered to us, through our Inner-Being. Our job, then, is to identify what experience we desire and focus diligently on it.

When beginning your day, start out by consciously deciding to notice the special people, places, and events that give you pleasure. Throughout the day, make a mental note of those experiences you enjoy the most. They can be anything or any combination of things.

For example: You may see a community leader at your office and want to be a leader yourself, but you do not have to take other aspects of their lifestyle that go with *their* vision. You can write your own script and fit it to *your* desires. Make a list of the things that make you happy and fine tune it as you go along. Feel the emotions as you visualize the new experiences. What kind of career do you see yourself in, what kind of marriage, what type of city do you live in—really feel yourself in these new experiences.

Spend some time each night until you have a nice script prepared. Do this to the point of actually seeing yourself in the role you have laid out. Now welcome your new prepaving thoughts;

they will replace your "old" prepaving thoughts or script. Whether you knew they were there or not, they were! Remember, you can't get what you want if you don't know what it is!

If you had trouble completing this script for yourself, you may have run into the second obstacle to identifying your heart's desire—the fear of *wanting*.

When our visions remain unfulfilled, we usually become depressed. There are two ways, however, to ease the pain:

1. Fulfill the vision—or get what we want.
2. Stop the vision by squelching desire.

Most of us stop the vision! Somewhere along the line we've come to believe that it's not acceptable to want a vision and realize it. Everyday we hear, "You can't do that" or, "It will never happen" or, "Things like that are not for you!"

Sometimes, it's easier to stop dreaming because it makes the short term pain go away. Unfortunately, the long term effects of a dream denied are worse.

We can actually become afraid of the wanting itself. We dread the pain it has brought us in the past, so we put a moratorium on visions for the future. W*e literally stop wanting.*

We become afraid to *want* happiness—we think we don't deserve it. We become afraid to *want* success—we think we might fail. We hesitate *wanting* a relationship—we think we might get hurt. Because of these past experiences, we stop the co-creating, and do not allow our Inner-Being to signal us with positive emotion. But at the same time, we are not having a lot of negative emotion either, so we don't do anything and let our lives become stagnant.

The key is to *know* what we want. As the saying goes, "Be careful what you wish for, you may just get it!"

Some of our worst experiences come from things we thought we wanted, only to find out later we would never want them again. But aren't those the first two Manifestation Procedures?

Visualize Yourself Already There

"Blessed are those that have not seen, yet believe."

-Jesus

We've operated under Universal Precepts since the beginning, with or without our knowledge. Through our drive and determination we have had great achievements on this planet. But with an understanding of how things work, we can do even more – and in less time.

Have you ever noticed how some people seem to get what they want, and they're able to do it in a relatively short amount of time? If you were to ask them how, their response may be, "I just knew I was going to do it!" or, "No one was going to stop me," or possibly, "I knew I could do it and I just kept believing in myself."

Are these high achievers blessed? The answer is, yes! They continued to believe in their vision even when they did not see it manifested immediately.

I believe this is what many Spiritual Masters have been expressing for thousands of years. How they lived their lives let people know that everyone had the very same power. Those who believed in their teachings were able to blend with that inner part of themselves.

Once you've identified what you want, the next step is to visualize yourself already there. Imagine the experience you would be having in your new environment. Do this to the point of actually feeling yourself there. See yourself there, and feel all of the emotions. (If you feel good when doing this, then your Inner-Being is signaling you with positive emotion.) *Do this everyday!*

Each day, take time away from the distractions in your life and be quiet. As you repeat this process, your Inner-Being will take aim at your vision, and this new experience will be on its way to you from the Universe! There are two very important parts to this procedure. *You must believe it is going to work, and, the higher your passion, the faster it will be delivered.*

The belief side of the formula is directly related to the visualization process as it adjusts your current vibrational state to that of your vision. Consequently, you will vibrate to the frequency required to attract the new vision. You must believe in your vision to the point of existent knowing. In other words, you must live today as though you are already having the experience in the vision. You must *feel* as though you are already there.

Once you have reached this knowing place, you will soon be paired with the experience and its associated conditions. The more childlike, unhindered emotion and passion you have, the sooner your vision will manifest on the physical plane.

"Truly I say to you, whoever does not receive the kingdom of God like a child, shall not enter it at all."

-Jesus

Belief and passion are actually co-facilitating the vision in a complementary fashion. Stated more simply: If you have a very strong belief in a certain lifestyle, even though your desire for it may not be galvanized, your strong belief will co-create the experience with the Universe. This is precisely how we maintain our everyday life. We *know* we will get up tomorrow, and we *know* we will drive to work in traffic. After that, we *will* enter the office building and we *will* do our job. This lifestyle is maintained, though our passion for it may be relatively low.

The belief side of the formula has been carried in many philosophies since the beginning of time. They advise us that a strong belief will help enhance life. Each of us has heard at some point or another that we must have hope, or faith in the future.

It is very important to be in harmony with our belief structures—whatever they may be—so we will be in a positive state of attraction. It is belief and passion that allows the Universe to fulfill our desired experiences, while it is the blending with our Inner-Being that will ensure happiness.

So our job up to now seems to be simple. Focus on the experience we want. Believe it is on the way with childlike passion. Allow the Universe to fulfill it, and co-create it with our Inner-Being. Not too difficult, right?

The next step will separate the dreamers from the achievers. We must fully maintain our current situational awareness, so we can recognize any related experiences that may occur in our day-to-day life.

Allow Your Environment To Align

The experiences we have in life occur through attraction. Several experiences, or events which appear related are sometimes called coincidental. These experiences bring vibrationally matched circumstances and conditions, thus creating our present environment, and reality. It is our level of situational awareness that determines whether we take advantage of these environments.

We've already discussed how we co-create and attract our personal realities. Now, some of us may be saying, "Hey, I never wanted that to happen in my life. I never focused on that experience! It just happened, maybe it was bad luck!"

When negative coincidences occur it is so easy to shrug off responsibility for them. It is much easier to blame it on bad luck. On the other hand, we tend to attribute positive coincidences to good luck. We sell ourselves short on both counts by *not taking responsibility.*

It is here that our situational awareness comes into play as our greatest ally. We must pay close attention to our current reality while our thoughts are focused on all that is joyful in it. The combination of having a vision of your future combined with looking for the good in your present life will allow your Inner-Being to deliver what it has targeted. Because you are looking for the good in your present reality, your emotional state will be more positive, thereby accelerating the delivery of your vision and future reality.

I would like to share an experience in which my environment aligned so quickly that I often recall it when I am a bit low on faith.

Though the majority of **Captain's Discretion** was written in the Georgia mountains, some of it had to be done on the road. I would take my laptop computer on my trips in the hopes of having some free time to work.

After a short writing session one evening, I was trying to decide if this book should lean towards philosophy, or "applied" philosophy, i.e., should it have a method? I just couldn't decide which way to go. I turned off my computer and went to bed, knowing that I would have the answer soon.

The next day I was traveling on a 727 as a passenger. I took my middle seat in the back of the plane, and realized the flight was filling up fast. Normally I would have tried to slide over to the aisle seat, but this time the plane was just too full.

I pulled out my notes on microtubules and began working. Within a few minutes, a pleasant, middle-aged married couple sat down on either side of me. I offered them the option of sitting together, which they gladly accepted.

During the subsequent seat shuffling, the husband noticed my notebook. He immediately exclaimed, "Are you writing about microtubules? We just came from a seminar relating to applied psychology and the microtubule function."

We proceeded to have a wonderful conversation through the entire flight, a most pleasant and enlightening exchange. From this discussion, I attracted other people and events, which persuaded me to include a method.

Some of us can remember being told as children, "Be happy with what you have." I would bet every time you heard that phrase, you felt negative emotion. That is probably because you also heard a hidden message along with it: "You cannot have your vision!"

But such negativity does not have to be true for you now. However, it is helpful to look for the positive in everyday life –in fact, it's even easier to do when you accept responsibility for your day-to-day environment.

How then, does our new environment align itself? Because your thoughts are beginning to vibrationally align themselves to your vision, they will begin to attract other vibrationally matched thoughts. As these grow and multiply, they will begin to attract other vibrationally synchronized people and circumstances. With these new people will come new ideas and new experiences. Finally, all of this will be in harmony with your vision.

As time goes on, your *entire existence* will be a new vibrational vignette of that vision. You will literally be a new person! This will go all the way down to that vibrational sub-atomic level we discussed earlier. It will affect your total reality.

For openers, you will feel emotionally more alive and have more energy. (Your energy, after all, comes from your Inner-Being.) You may experience what I call vibrational anticipation, or a childlike excitement. During this stage it is essential to keep focused and get ready to take action. This vibrational anticipation is the energy you need to put your newly attracted environment in order. Do not direct this high energy into non-action, or it may become vibrational "anxiety." Now that everything you need for your vision is being delivered, you must take the *time* to put it together. Time is that aspect of physical existence that allows us to *feel the contrast* of manifesting our dreams. This truly is the fun part, the creating itself combined with vibrational anticipation. When we put our attracted experiences together, with this internal energy, we may find this to be more enjoyable than simply observing the completed vision.

The more complex the vision, the longer the alignment process will take. This is because we tend to take *more time* when we make dramatic vibrational state changes.

For example, I had a vision of publishing a few articles in an Aviation Human Factors quarterly. Since I was in the business, it only took a few weeks for someone to ask me for an article. Most of my environment had already aligned before I even thought of writing anything. All of this happened relatively quickly as I was already at the vibrational state required. My desired experience did not require any *new* experiences to change my vibrational state.

When I decided, however, to initiate a career in writing and seminar production, I found I needed more time. As I visualized my new experience, I noticed a shifting in my environment, but not in a manner I could have planned.

Some of the new experiences were absolutely wonderful: I thoroughly enjoyed working with the people in NASA Human Factors. It was great to exchange ideas with the training specialists from Boeing – I always enjoyed facilitating classes for my fellow pilots at the airline. It was fascinating to share an evening with a Shaman under a midnight sky in Chiclayo, Peru. It was exhilarating to meet a publisher, editor, and marketing person at "just the right moment." These new people and events all made sense. I could see they were connected to my vision and it was easy to accept them!

There was one experience, however, that I initially had trouble understanding. I was offered my opportunity to be based in another city with my airline. It felt good to accept. The move seemed in tune with my long term visions. So I took the offer–it was to be for one year.

After the first few months, I began to wonder if I'd made the right decision. Although I still felt I was on my vision path, I could not understand why here: I was separated from my children most of the time, and I had no friends. This particular city's weather was cloudy and rainy much of the time. I did not get to fly that often, but rather, had to standby in a hotel 24 hours a day, 21 days a month. I seemed to be making little progress towards my vision. What was I doing here? How did it relate to anything I was visualizing? I felt neither like a pilot, nor a writer. *Who was* I?

I continued my morning meditations, trying to stay centered on the vision. I focused on everything positive in my environment to keep me at a positive point of attraction.

Soon the fog began to lift and I discovered that some of the new base instructors, were very interested in the research I had done, and the material in **Captain's Discretion**.

Before long, we were fine-tuning my seminar presentations, sharing many related experiences, and enhancing all of our visions. You see, these people were waiting for this information also. It was truly a win-win scenario.

So as our environment aligns during complex visions, we may have new experiences that will seem unrelated. We may even feel a loss of identity, which is exactly what is happening. We are changing identities or vibrational states. I call this phase "limbo" because that's exactly what it feels like. We are not the *old* us, and we are not yet the *new* us. We are in transition, and we may feel lost.

> *" 'I am,' I said*
> *To no one there*
> *And no one heard at all*
> *Not even the chair*
> *'I am,' I cried*
> *'I am,' said I*
> *And I am lost, and I can't*
> *even say why*
> *Leavin' me lonely still..."*
>
> *-Neil Diamond* [1]

This is where most people lose faith and abandon their dreams. Because we are not receiving the full benefits of the new experience, and have stopped attracting some of the benefits of our previous state, we may want to give up and try to return to our old familiar state. If we drop our dream here, that which we really wanted, after having partially changed our vibrational state, we may find that returning to our old ways does not bring us happiness. That's why it is so important to remain focused on our vision while *passing through* the limbo stage.

If we have a powerful vision to carry us through new life experiences, these experiences will eventually make sense. We will understand why we had them and they will add value to our life, rather than confusion.

In addition, I also noticed something during my limbo transition. I constantly had an underlying sense of happiness through it all, because I allowed my Inner-Being to guide me. Though I found my desired vision taking me through unknown new experiences, all I had to do was trust and allow, thereby keeping my thoughts focused on my vision. My environment aligned perfectly despite my not being able to understand it at the time.

It was also at this time that I was able to genuinely share my deepest beliefs with my children. Though we saw each other on a less than regular basis during most of that year, they went through the whole creative process with me. They saw the rough drafts, met my publisher, and felt the vision. They learned faith, patience, and trust right along with me. As I explained why we were separated during those difficult months, I found it natural and timely to share my way of living. It was real, not just a lesson at school; they were living it, experiencing it with me.

In short, when we allow our environment to align during complex visions, we must allow new experiences to carry us there. If we listen to our Inner-Being, through our emotions, we will feel joy as we keep our thoughts centered on our vision. If we focus on the apparent negative aspects of our environment during the alignment, we may lose sight of the vision and put ourselves in a negative point of attraction, which will only attract more of the same. Instead, we must remain focused during the limbo transition and stay in a positive state of attraction.

We lose our dreams and visions when we do not recognize the environment we attracted to help create them.

Physically Take Appropriate Action.

We live in a space-time continuum. This is our *world* as we know it. Everything we are aware of on the physical plane is an expression of Energy, in the form of creations throughout the Universe. This focused Energy is also the same Energy that created us, the physical human being. We exist in *time* as long as we have a physical body and *perspective* from this dimension or *space*. We can only have this physical existence while in this time. Again, our perspective as human beings is based on this physical space-time existence, however, *it is not limited to it.*

We, the physical beings, have been given a great gift: The opportunity to incarnate into the physical, and blend with the spiritual, to facilitate co-creating with the Universe. This is what creates purpose in our lives.

Up to now, many of us have been randomly directing our energy—randomly co-creating, because of a lack of focus. As a result, we may get the *feeling* of lack of purpose. Focus and direction, in tune with the Energy flow, allows us the feeling of purpose.

The Earth, the sphere itself, is a complex, self-balancing, creation in its own right, a host to many living creatures. It provides an environment, a place to physically incarnate so that we may co-create as a unique expression with purpose. The more we are in tune with the earth's environment, the easier for us to adapt to physical life. There are many benefits that come with earthly incarnation, but we shall concentrate primarily on two: *contrast and time.*

When we have a physical body, we have the contrast of existence that goes along with it. We get the ability to determine contrast through our six senses (yes, six): Touch, smell, taste, sight, hearing, and emotion. It is this last one, I want to talk about.

Our emotion is the best guidance system we have—we should not try to repress or ignore it. If we pay attention, we will value it as one of our good senses. With this new tool, our joyous co-creating can begin.

After your environment has aligned and you've not only attracted the components of your vision, but have recognized them as well, it is time to put it all together. As we've stated before, time is our ally which we can utilize to physically focus, through our Inner-Being, the power of the Universe.

Since the space-time continuum is full of contrast, we are capable of co-creating a vision with our Inner-Being that will bring us much joy through our emotions—a joy that will make us happy, as well as energize us while we are taking the action to physically create our vision. Our physical work will be action, galvanized by joy!

There is a big difference between this kind of action and just working hard. I'm sure you know a lot of people who *really do* work hard, and yet never seem to get anywhere. It could be because they are not tapped into the power of the Universe, not in sync with the vision. We have all tried to muscle things into a vision, only to find ourselves exhausted in the process.

The key here is to monitor your emotions, your inner guidance, making certain that you're meshing your physical attractions the way you need. In fact, you will probably find the physical action to be the easiest of the five procedures, and also the most fun.

If you've noticed, when I talk about physically taking action, I am mostly concerned with time. I truly believe time is the key to action. We have to discipline ourselves to take the time to manifest our dreams. For example, let's say you want to be an airline pilot. This is your vision. But you are currently in high school, have no money, wear glasses, and have never been in a plane. Your guidance counselor tells you flight school is very expensive, unless you go the military route. Of course, they won't accept you because you wear glasses. Where do you go from here? Well....

The first thing is to *know* that you are going to be an airline pilot. See yourself already there. *Feel the experience.* Visualize the career as best you can. Imagine sitting behind the controls of a big 727, just like in the movies. At school you will bump into other, more encouraging teachers who just happen to know a friend who had glasses and went on to pilot training. Before you know it, you will also be applying to this particular university, and because it is fully accredited, you can get loans. You will have to work at two or three jobs on the side, but you won't care: Your dream is coming true! Oh, but what about Mom, she wouldn't want you to leave the safety of Wisconsin, and go off to some flying school in Daytona Beach! How will you ever convince her? To your surprise she thinks it's a great idea. Dad notices the name of the university and exclaims, "Hey, I went to this college during World War ll for engineering. Great school!" Next thing you know, you're on a big jet, (you have no idea which type) heading to school to learn how to fly. You ponder – "How could all this happen so fast?"

As you get off the plane with childlike excitement, you stick your head into the cockpit and say, "Thanks for the good flight, gentlemen. I'm starting flying lessons this week and someday I'll be sitting up here, too!" While you're waving good-bye, the Flight Engineer looks right at you and says, "Not with those glasses."

Well, I'm still wearing glasses, and there are still people who think negatively. When I started all this, I didn't know what Precepts or vibrational vignettes were. I simply *knew* I was going to be an airline pilot.

So the school was delivered to me, as were some tutoring jobs. I borrowed every nickel I could get my hands on and began flight instructing my Junior year. I studied every night, always focusing on flying. This was *my* part of the deal. I had to take the *time* daily to absorb information. Soon it became my information. Some students were faster at this than I was, some were slower. I practiced cockpit skills in the University's aircraft over and over again— just like the curriculum dictated. Sometimes I needed more time than the course allowed; sometimes I needed less.

I now know practice is actually repeated focus of energy into the physical body itself. In other words, the body will change its vibrational state over time with repeated practice at something. The time required will decrease with stronger focus or energy, which is why concentration is so important in sports. It also explains the muscle memory needed for playing the piano or even flow checks used in cockpits. Our bodies know where they are, and they are listening for direction.

12

Programming Your
Internal Reference System

Modern jet airliners navigate using Inertial Reference Systems (IRS). These navigation systems are capable of guiding an aircraft along its intended route of flight with very accurate precision. They are comprised of computers, lazerbeam gyroscopic comparators, and visual displays for the pilot, all working in harmony to let the aircraft know its location at any given time. It is capable of providing this data without any external navigation signals. It will always know where it is, assuming it is *programmed correctly.*

There are several steps to programming the IRS, but they can be classified into three main procedures:

1. Enter current position.
2. Enter final destination.
3. Enter intended route of flight.

The computer must know the aircraft's current position, before anything else can happen. The pilot must program the exact latitude and longitude into the system, thus enabling the aircraft to compute any motion from that starting point. When this is done correctly, the laser beam gyros will then be able to compute the slightest movement. If the pilot enters an incorrect starting point, by missing one digit in the latitude or longitude, the aircraft will actually think it is at a false location and begin its computations. This is why at least two pilots always double check all programming.

Several years ago, however, there was a Korean 747 shot down over Soviet airspace because it was off course. The strongest theory to date suggests that the IRS computers were programmed incorrectly.

The next step to programming the IRS computer is to enter the destination. Again, the computer will need accurate information.

Next the pilot will enter the intended route of flight. The word *intended* is used, as there may be some minor adjustments along the way, and the route may have to be updated. Being flexible, while still accomplishing the mission, is an important trait of a good captain.

The route of flight is *how* the plane is going to get to its destination. As it is programmed, fuel burn is estimated, as well as time/distance requirements for the various checkpoints that will be passed. The pilot wants to be sure there is enough fuel on board to make the intended destination. Even though the winds and weather have been forcasted to be of a certain value, *conditions can change.* By utilizing situational awareness, the flight experience can still be a good one, with everyone arriving safe and sound at the desired destination.

The same holds true in life:

1. We must know where we are and what we are doing here.
2. We must know where we want to go and *what we want* in life.
3. We must decide how we will get there, or what roles we will take on to get us there.

If we are in a state of situational awareness, we will know our current latitude and longitude, or place of attraction, from where we are beginning our journey. Armed with this accurate information, we can begin to program our personal Internal Reference System.

From here we can begin prepaving our future and choosing happy destinations. We can begin to decide what type of experiences we want in life. The personal mission statement is one of the best tools to keep us on course.

A strong, clean mission statement is indispensable to creating the future you want. Not only is it a written confirmation of your future life experiences, fundamental values, and projected lifestyle, it is much more. *It is the first manifestation of your spiritual desires, as seen by your physical body.* This is very powerful!

When we put something in writing, we are able to add another one of our senses to the visualization—the sense of sight. This will enhance our emotional guidance system, because of more input into our vision.

The personal mission statement needs to include some basic and fundamental aspects, but it is mostly a reflection of your future desired life experiences.

You may want to begin your statement with several sentences describing your fundamental values and beliefs about yourself. Some examples are: honesty, integrity, balance, faith-centered, love, and compassion.

You may also decide that you want to continue to blend with your Inner-Being, and thus better utilize your natural guidance system. It is really a good idea to spend some time thinking over what the *core of your future experiences* will be attracted to.

One example could be as follows:

Through effective integration of body, mind, and Spirit, I will be happy. As I blend with my Inner-Being , I will listen to my emotions, and visualize with humility, understanding, and wisdom. Radiating compassion and love, I will be a positive influence in this world.

After this opening statement, list the roles you would need to assume to carry out your personal mission.

Every person's roles are different, ranging from parent, to employee, to teacher, to friend. The list never ends. We are all expressing ourselves in our own unique way. The world, happily, needs all of us.

Your personal mission statement needs to be put where you can read it everyday. I put mine on my kitchen wall, and another one in my bedroom night table. I also keep one in my suitcase for inspiration when I'm on the road.

Your personal mission statement is a visual reminder of what is most important to you. When you deal with a welter of daily distractions, it will serve as a reminder of what you really want. It will keep you on course. True, you'll have to update your flight plan from time to time, but it will always reflect your core values and direct you towards your destination.

If you have a family, you may want to have a family mission statement as well. It can be very helpful in creating a joint vision between two or more people.

With a family mission statement, the important thing is for everyone involved to have a part in creating it to help ensure understanding and enrollment. When someone buys into a family or joint vision, of any kind, it will naturally become part of their personal vision, and they will attract on an *individual* basis, feeding into the synergy of the *whole*.

Because the entire family will be involved, a family mission will have to be agreed upon and understood by those involved, including the children. The opening statement may reflect some of each person's personal mission statement, culminating into a core belief structure and set of values centered around a family environment of mutual respect.

Roles will have to be agreed upon to avoid confusion later. Each person must remain an individual while achieving the collective mission.

13

Responsibility And Power

According to *Federal Aviation Regulation* 91.3, "The pilot in command is directly *responsible* for, and is the final *authority* as to, the operation of the aircraft."

What I've always found fascinating about this definition is that the FAA clearly places responsibility before authority. The captain must first accept full responsibility for the operation of an aircraft–then he or she will be empowered with the authority to command it.

When we accept full responsibility for our life experiences, then and only then, will we be empowered to be our own Captain.

We're not talking about a limited concept here, but one which extends into all areas of our life: marriage, career, family, social circles, health, spiritual growth, and emotional well-being.

Once we accept responsibility, we can stop blaming other people or circumstances for our experiences. We can even stop blaming ourselves.

Blame and responsibility are two different things.

Blame: To accuse of being at fault; to put the responsibility of error on, to condemn.

Responsibility: Of being responsible; having accountability, being the source or cause of something.

Blame tends to suggest that someone has committed an error and should be condemned for it.

Responsibility leans more towards simple acknowledgment of *who* is behind it all, or *who* has the power to manifest a particular outcome. This type of responsibility is one of recognition and empowerment, not one of blame and condemnation.

When we have unpleasant experiences in life, we can fall victim to blaming another. We can attack the qualities in someone else whom we perceive to be inadequate, or who *makes us* unhappy. When we get into the blame mode, we give these same people the responsibility and its empowering authority over our lives. We hand over exactly what we seek—control and authority in our own life.

Sometimes, it is very difficult to accept responsibility for the varied experiences in which we find ourselves. We can get caught up in the blame game, and even dig into our past, looking for solutions or root causes. Though this may be helpful in discovering contributing factors to our current behavior or predicament, as long as we continue looking outside ourselves, we will never find what we are looking for: the ability to attract our desired future experiences from within.

> *"First of all, let me say that effective healing*
> *is a gift of the Holy Spirit.*
> *Therapists,if they are good, stop trying to fix their patients.*
> *They work primarily on the relationship,*
> *building a community of two."*
>
> -Dr. M. *Scott Peck* [2]

Some psychotherapeutic techniques employed today are not

able to finish the job they set out to do, but do offer a good starting point. They get their readers (or patients) into gut-wrenching sessions focused on *something they do not want in their lives*. In other words, they explore the negative experiences.

The theory being, with time, and the utilization of some self-discovery techniques, the patient may be able to confront these issues. This all leads to a change in the thinking process. The downside, however, is that this may lead to wallowing in these negative states for extended periods of time. It may be difficult, if not impossible, to move beyond something about which you are constantly reminded.

Much of our modern therapy is designed around the above premise. But where does it leave us? Do we feel good when we discover that we married someone just like our parents? Do we feel a sense of satisfaction when we recognize that our childhood was painful? Though all this information may be helpful in finding root causes, what do we do now?

Here is one family's story:

Nick and Brenda are beginning their fifteenth year of marriage. Brenda is a good homemaker and Nick is a good provider. They have two children – Scott, 11, and Susan, 8. Because Nick is an airline pilot, he is on the road about 18 days a month.

He and Brenda were both young when they took their vows. Nick, 22, was just starting his aviation career, and Brenda, 19, was still a sophomore in college with no definite career plans. Within a few years, Brenda got her degree, and Nick, one of the lucky ones, landed a job with a major airline. Then little Scott came along. Because of Nick's flying schedule, it seemed only logical that Brenda would stay home with the baby. She agreed, and before long, Katie made her entrance into their lives.

It is now ten years later and the children are in grade school. Having just been upgraded from First Officer to Captain, Nick is feeling justifiably proud of his accomplishment. Brenda tries to share Nick's enthusiasm but finds it difficult. There is even a feeling of envy on her part. After all, she's had no time for a career of her own, while tending to the kids and taking care of Nick's needs.

In the ensuing years, Brenda's resentment grows, manifesting itself gradually in her life. At first, a mild depression begins to surface, alternating with a brusque, sarcastic demeanor. Brenda's mood swings trigger a response in Nick to *help* his wife. To his dismay, he is met with increasing anger every time he tries to intervene.

The accusations begin flying, fast and furious:

"You don't appreciate me!"

"Stop telling me what to do!"

"You're domineering and overbearing!"

"You always think you're right!"

"You have prevented me from having a career!"

Nick is astonished and taken aback by Brenda's accusations. After a while, he begins responding in defensive anger.

"You said you wanted a family!"

"You could've started a career anytime you wanted – all you had to do was say so!"

"I thought you wanted to stay home!"

"I always did what I thought was *best* for us!"

Determined to hold the family together, Brenda elects counseling. Through psychotherapy, she discovers her *psychological* reasons for being where she is today. Eventually, the counselor is able to convey through self discovery, that Brenda is responsible for choosing Nick as her husband. She chose him because he had many of the personality characteristics and a similar psychological makeup as her father: domineering, strong-willed, overbearing, to name a few of the negative traits. On the other hand, he is also highly motivated, decisive, and has a sense of humor.

The therapist tells Brenda that one reason she was attracted to Nick was in order to work out her inner "psychological child-like profile." She now begins examining her entire life, including her childhood. It is not pleasant.

Brenda soon feels that she was, and still is, somewhat of a *victim* of these other people's inadequacies. In addition, she begins to learn that while she cannot control Nick, she can choose how to *respond* to him.

Armed with this new sense of *response control*, Brenda's self-esteem rises, enabling a new found power to manifest within. Before long, she asks Nick to attend marriage counseling with her. Although he does not enjoy discussing personal problems with strangers, Nick does attend. Like Brenda, he wants to save the marriage.

The joint counseling sessions continue for another two years. Nick and Brenda learn some communication skills and seem to be able to cohabitate together. Every now and then, they even have fun. After all, they do love each other.

Brenda still resents Nick for being overbearing, though she seems to be able to *tolerate* it for now. She cuts down on visits to her father, fearful that an argument may result. Her father has been *right* for 60 years.

As for her mother, Brenda harbors secret feelings: a mix of love, pity, and anger. She wishes her mother could have stood up to her father, but at the same time understands why she could not.

Who's right? Who's wrong?

Unfortunately, this situation is all too common: *Two people trying to do what is right, yet both are in pain.*

In order to understand how we attract similar experiences into our lives, we must first know who we are, and how we fit into the Universe. We must blend with our Inner-Being if we are to utilize that always dependable internal guidance system. To be the captain of our lives, we must empower ourselves with the *responsibility of our thoughts and situational awareness of our vibrational state of existence.*

Once we know that we are much more than we can see, we will be able to direct more spiritual energy into areas of our life that we want to flourish. We will be guided through experiences, ensuring our happiness. Through our focused thoughts we will be able to do things we never before imagined possible.

III

BALANCE AND SELF MASTERY:
THE SEASONED CAPTAIN

14

The Seasoned Captain

When an airline captain is affectionately recognized as a "seasoned" pilot, it is because he or she has accumulated wisdom and knowledge commensurate with numerous and varied flying experiences. Through the years, they have learned the limitations of aircraft, flight crews, air traffic control, meteorologists, and countless other critical aspects of their environments. They have been to the "edge of the envelope" with all of them, in the process developing their own patterns of integration that continually serve them well. In essence, they have actually discovered their *own limitations* with regard to these variables, and thereby created their *own* balance with them. As they fly around the world, that internal balance will give them the adaptability to command effectively, as those same variables configure into different scenarios.

If we want balance in our lives, we must first know our *current* personal limitations. This knowledge will help us maintain balance, which in turn will help keep us in a state of positive attraction. From this position of strength, we will have the energy to expand those current limitations or "make the envelope bigger!" And that's what most of us came here to do!

When we undertake a big project, we usually find it more manageable if it's divided into smaller tasks. By so doing, each area gets the attention (energy) it requires, while the entire project is still being served. And what project could be more magnificent than our entire life!

We can divide our life into five major categories:

1. **Faith**
2. **Family / Social**
3. **Finance**
4. **Fitness**
5. **Fun**

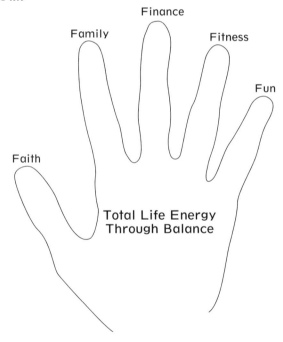

In the interest of memory recall, they all start with the letter "F". In addition there are five fingers on a hand, so this will help us remember the five areas of our life needing attention. We can sometimes unknowingly ignore one or more areas, only to find out later that our life is out of balance, leading to *total energy* reduction. The total energy flow is what's important here, but we must understand each area before we comprehend the whole.

Faith

When we have faith, or choose to believe that our future, co-created vision of ourselves is being delivered to us, we are in a positive position to attract it. This is very important for self-mastery, which, in turn, will create balance in our lives. This knowing and its resultant *feeling* is the catalyst of it all.

As we apply the "Five Manifestation Procedures" to each of these areas in our life, we will gain control over our experiences, resulting in an increase in total energy of a self perpetuating kind. It is when we do not have faith, or knowing, that we get out of balance. I call it "Reactionary energy."

Reactionary energy is the energy we spend on those day-to-day experiences we do not want. When we do not prepave our experiences with a personal vision, and live by default or chance, we have to react to our environment, just to get by. *We already know reacting takes an enormous amount of energy!* Naturally, we will have difficulty finding balance as we will have little energy remaining for other areas of our life.

Our internal faith can also be incorporated into our religious beliefs, and from there we can radiate it out into the community. It is a method that works much better, I think, than trying to take our faith *from* the religion.

Family/Social

Though situational awareness is a great asset, it can still elude us. When we have family or friends around us, they can help maintain our balance by reflecting back to us, in a loving manner, who we are and where we're going. It can be a thoroughly satisfying interaction capable of planting the seeds for our future growth.

As you may recall from *Precept Number* 2, the people most dear to us will be vibrating at a similar frequency. As a result, we will see in them the potential for their growth, as well as the yearning for our own growth. The vibrational shifts that are desired, and required, will probably be very similar. Loving, caring, and allowing relationships in this area can supply unlimited energy. When two people recognize one another's positive potential, their simple presence (way of living) can function as a catalyst in opening each other's energy flow.

The same holds true when children are involved. Their innocent, childlike zest for life can remind us of who we want to be. And while we are trying to provide them with direction in life, we will sometimes find our own path taking a few surprising turns.

All of these relationships provide a marvelous sounding board for our vibrational vignettes. And while we are bouncing back and forth in what may appear to be a state of confusion, we are really discovering our current limitations which, of course, will help create balance. And balance is exactly the position of strength we seek.

Finance

Because we all need financial resources in the physical world, most of us have some type of job or career. Invariably, we are providing goods or services to other people. The financial rewards we reap, however, are directly related to the level of attraction others have for our product or service. The amount of joy we receive from our career is, finally, up to us.

We have assigned many powerful values to money, some good, some bad—but in the end, we may utilize money in any way we choose. The earlier in life we discover this truth, the sooner we will be able to integrate it.

Before we discuss how all of us attract different finances, let's take a look at the difference between attracted conditions and attracted experiences.

Attracted Conditions: Those aspects in one's life which makeup the physical environment. Neutral by nature, they may, when interacted with, create varied experiences. Conditions can be attracted by themselves or as part of an overall experience.

Attracted Experiences: The integration of one's life with those conditions in it, resulting in positive or negative emotion. Attracted experiences bring with them associated conditions. It is best to visualize experiences because your Inner-Being will deliver the optimum conditions to create the positive emotion.

When we visualize future experiences for ourselves, we are also attracting a financial aspect as well, whether we know it or not. Money is simply one of the associated conditions delivered with our experiences. Again, our perception of an experience with money has a great effect on the amount we will attract. There are two main inhibitors, however, to our actually getting the financial benefits so many of us seek.

Focusing on our lack of finances: When asked, most people will admit to wanting a higher salary, but will dwell instead on their current reality of having a *lack* of money. Many of us see ourselves in the future as we are now, just getting by. In other words, when you hear the word money, are your first thoughts "plenty" or "not enough"? Remember, these are your prepaving thoughts.

Money is the root of all evil paradigm: It is true that some high earners have reaped obvious hardships in various areas of their lives. But that can happen to anyone, rich or poor, who gets out of balance. More often than not, people lose different aspects of their lives – even money itself–because they failed to understand how high status was attracted in the first place. In other words, we become addicted to a certain life condition, which, in turn, generates a fear of immediate loss – which then manifests itself into reality through prepaving thoughts. The key here is to visualize yourself having a joyful experience in the new lifestyle. Encompass the other important areas with your vision, i.e., family, health, faith, etc.

When we visualize the experience, the Universe will deliver the conditions to keep us in balance, including the right financial conditions.

Fitness

We manifest into the physical world through the expression of our physical body. Ironically, our spiritually based Inner-Being has more potential influence on the condition of our bodies than it does on any other aspect of our life experience. Understanding the relationship between our physical body and our Inner-Being, or our "physio-spiritual" makeup, will serve us well. In fact, directing energy into this critical area early on can be a great catalyst for increasing energy into all five areas.

The first step in creating the physical body you want is to see it *before* you even lift the first weight, or walk the first mile. Remember, you are co-creating an experience for yourself in the future. And this experience must be one that you want to happen, one you believe will happen.

Visualize the body you will have in a few weeks, in a few months, in a year from now. Every time you look in the mirror, see yourself as you want to be; *think* of yourself as at that weight or in that condition. Say things such as, "Hey, I look good. I'm physically fit, I feel good and I'm balanced" versus saying, "Gee, I look like a beached whale, but someday I'll be in shape."

With daily practice, your vision (of yourself) will be on the way, and your environment will realign itself with the enabling power of the Universe behind it. As you start to vibrate at your vision's frequency, you will harmonize with the conditions that help develop a good body, e.g., comfortable exercise and enjoyable diet. This new regime will feel good to you, because you are now at a new frequency. To your surprise, you may find yourself enjoying food and exercise easier to accomplish than you ever imagined. (I purposely use the word "enjoy" when referencing food, because I believe the pleasure of eating increases as balance is reached.)

Does this mean all of us should aspire to the front cover of "Muscles Are Me" magazine? Absolutely not! The goal of our health plan is to create equilibrium in our physio-spiritual relationship. Let me explain.

Our Inner-Being signals us with positive emotion when we bring our thoughts into harmony with our overall greater intent. The physical body is our "vehicle" to carry out these intentions, having been created in our Inner-Being's image. Each of our bodies is as unique as the intentions of our Inner-Being. So, our job is to keep that body fit enough to carry out those intentions – all of which creates a state of equilibrium in our physio-spiritual makeup. We will know when we are at that point of balance by monitoring our emotions, or inner guidance system. We will actually be able to *feel* when our body is in the required shape, to handle our future visions. This should be our standard of measurement, not brushed photographs of professional models. Remember, their life's vision is as unique to them as yours!

When we live near our equilibrium point, not only will our physical energy increase (because our body requires less energy for basic maintenance), but our total life energy will also increase. Our energy valve will open and pump more energy into all areas of our life. This increase in Energy flow will happen because we have achieved the balance necessary to realize our visions. Finance, fun, family, faith, and fitness, will all receive more energy, thereby keeping us in a state of positive attraction.

The reason physio-spiritual equilibrium is more desirable than just an overly aggressive work out, is the balance it creates with other areas of our life. Although the endorphins released in our body during exercise make us feel good, they are actually *pain killers*, designed to allow us to continue exerting physical energy longer than we normally would. So we do not want to work out to the point of becoming addicted to the endorphins – we would lose situational awareness. Should that occur, our total energy would decrease, sharply signaling us with negative emotion. At this point, we need to examine the balance in our life before some form of crisis occurs, such as sickness, injury, depression, etc.

To sum up: we can envision that physio-spiritual equilibrium that is unique to each one of us. We will know when we are there by monitoring our emotions. Likewise, a good fitness program complements other related areas of life, but should not be used as a long term escape from areas needing attention. As we discover our *current* limitations, we will begin moving towards balance, which will increase our total energy flow, which in turn, will increase all of those same limits, thereby expanding the edge of our *total* envelope!

Fun

Fun keeps us in balance by opening our energy valve while we allow our new environments to align. Because we live in space-time, everything doesn't always happen right away. If we allow ourselves this enjoyment, we will be in a more positive state of attraction, thereby speeding up the process.

Fun also makes it much easier to find those things in your life that you can appreciate right now, in your current environment. This not only adds to your current joy, but again, it adds emotion (speed) to your future vision.

Hey, you might as well have fun at this!

As we can see, all five fingers must be balanced, and flexible, to maintain a strong *total energy flow*. If one area, or finger, becomes weak or out of balance, the hand, or our life, will experience a decrease in total energy.

15

And The Married Couples?

A marriage is the sharing of a joint vision between two loving people. This joint vision expresses the power of the Universe, equal only to the energy of the personal vision of each partner. In other words, each partner must have their own vision, then they will have something to give to the common ground. In order for both to attract personally inspired life experiences, they must remain in positive emotion. Their inner guidance signal tells them when they are connected to the flow of positive energy. If either partner judges, or enters the blame game, they will cut off the flow of energy to *themselves*. This will resultantly cut off the energy to the joint vision. Negative emotion will follow, and the downhill cycle begins. The other partner usually follows suit.

Of all the many different reasons for falling in love, an obvious one has to do with our childhood resonance. If our thought patterns are basically the same as they were during childhood, we still may be resonating at that earlier frequency. Should this be the case, we will probably attract people similar to one or both of our parents. This is where *pre*-marital counseling can be utilized to increase situational awareness.

Another way to avoid getting into an unhappy marriage is making use of personal and family mission statements. If both people know where they are going as individuals, they will have a better chance of deciding where to go as a couple.

16

How Do I Visualize Relationships?

Several years ago, I had a picture in my mind of the woman I wanted to meet. My ideal mate was intelligent, kind, and sincere. She had long hair tied back in a pony tail, wore very little make-up, and dressed in blue jeans. She enjoyed being outdoors and loved horses. In addition, she was European and in her thirties. Thinking this was a reasonable vision for a young man to have, I looked and waited.

To my delight, I was introduced to a beautiful, charming woman at a dinner party—let's call her Beth. Beth was about five feet seven inches tall, had long hair which she kept in a pony tail, and hardly wore any makeup. When I met her, she was wearing blues jeans with a denim jacket—suitable attire for an outdoors woman who worked with horses. In fact, many of these horses came from Germany, where she was born.

Since we met at a social gathering, we had enough time to get acquainted. Beth was a great conversationalist and we enjoyed each other's company. After dinner, I said goodnight to her.

I never saw her again. The opportunity came and went, and that was it. I had met this ideal person, but nothing happened. I began to ask myself what I really wanted in a relationship. The truth was, I did not know.

I hadn't visualized an *experience* of having a relationship. I had only visualized the *conditions* of this other person. I was short-circuiting the Universe by not visualizing my interaction or experience with this ideal person. In other words, I did not visualize how I *would* feel when meeting her. If I had, I would've attracted someone to match that emotion during that brief but wonderful experience. But because of my own internal confusion, it was very unlikely that anything could possibly manifest resulting in happiness.

So how, then, do we visualize a relationship?

We do it the same way as we co-create our future reality. We visualize the experience we want to have and feel! We cannot, of course, visualize the behavior of another who is in control of his or her own life. We must allow our vibrational state to attract the person who will match the emotions we feel during our visualization. Stated another way, the Universe will present us with the appropriate person to give us that experience. Because a relationship takes two people, the person you attract will also have attracted you.

Let's run through an example using the five Manifestation Procedures.

Step one: Be aware of what you do not want.

Make a list of experiences you want to avoid with your future mate and feel them as you write them down. By doing so you will help fine-tune your situational awareness in recognizing (but not focusing on) bad past relationships which you do not want to repeat. For example, ask yourself if you feel manipulated, frustrated, angry, sad, etc. Remember to note how *you* feel, not how the other person acts.

Step two: Identify that which you do want.

Make a list of experiences you want to have with your future mate. Feel the emotions; visualize yourself sharing the day with them. When you wake up in the morning, do you have coffee, or go jogging together? Whatever it is, feel the emotion of sharing. As you track your emotions during the day, feel yourself sharing with this other happy person. Pay attention to *your* emotions during the visualization and keep adding experiences to the vision that feel good. You will soon be vibrating at this new frequency.

Do not focus too intensely on the physical appearance of the other person. (This can distract you from *your* emotional feeling.) Simply see yourself happy when you are with him or her, thereby allowing the Universe to present someone with whom you will be happy in all respects, naturally including physical appearance.

But isn't this very one-sided? After all, what about the other person?

Like a good marriage, a healthy relationship is a joint vision between two people who have strong personal visions. Your Inner-Being will not attract someone incompatible with your vibrational state. On the contrary, you will match up perfectly. The experience will be happy because both of you are attracting the same thing.

Relationships are especially tricky. If we don't know what kind of experiences we want in life in general, we may have trouble attracting the mate we seek. Many of us find ourselves in relationships which start off wonderfully, only then to turn unpleasant.

What accounts for this phenomenon?

More often than not, we attract people who vibrate at a frequency we consciously do not choose. Which is why it's so important to have a healthy vision of our own life before we try to involve all of the aspects of another person's. We are always attracting, whether we know it or not.

Current psychology offers us many ways to discover thought patterns within ourselves, that do not promote healthy relationships. It is possible that one of these methods may be useful in helping us find what we *do not want*. The majority of unwanted attractions, however, are the result of past experiences in which we have locked into some type of thought pattern that determines our point of attraction. These old patterns can be from our childhood, or maybe even from a series of adult relationships.

We must recognize that new relationships begin forming before we ever meet the other person. If we truly want to have the relationship of our dreams, we must first be the person of our dreams.

Step three: Visualize yourself already there.

Visualize yourself in the relationship. Feel the emotions as you picture yourself sharing the experience with the other person. Combine your desired experiences with this new person in your visualizations. Situational awareness combined with a focused, happy vision of your future will attract someone who is very compatible.

Step four: Allow your environment to align.

Pay attention to gut feelings while in a state of situational awareness. When you meet the person who matches your vibrational level, you will be attracted to each other.

Step Five: Physically take appropriate action.

It is now time to start enjoying the relationship, while monitoring your inner guidance.

When two people enter into a loving relationship, they naturally become aware of three separate visions: their own personal vision, the other's, and their joint vision. The joint vision may be one in which both parties are helping each other co-create a personal vision. This is a most desirable state, as now two people can have synergy, or an increase in total energy flowing between them because of their mutual support.

17

When The Medical Doctor Can't Help

What do you do when a trusted doctor, tells you that you may go blind in one eye—and there is nothing he can do? Well, if I had heard this several years ago, I would have panicked. Fortunately, the doctor said it several years later.

When I was in my twenties, and just starting my airline career, I developed an eye disease called Macular Degeneration. It is a disease which leads to loss of vision, and often to blindness. I have had it twice in my life.

In some instances, the Macular Degeneration can be corrected with laser surgery, as was the case for me many years ago. The doctor directed the laser to the exact point of degeneration, while not doing any collateral damage to the remaining retina. Within a few days, my vision returned to normal and I was on my way.

But the second time was different. The visual deterioration was just as noticeable, and the medical letters to the FAA had to be sent out—only this time the doctor offered a different diagnosis. He announced that he could not use a laser to repair the

Macula. If he did, the center of vision would be destroyed, resulting in blindness.

I returned home and immediately began meditating, obviously concerned with the state of my eye. I began asking myself some hard questions. Within five days my vision returned to normal.

The following paragraphs summarize the results of my personal experience with this physical condition I attracted. I have organized this episode into a chapter. It was designed during my experience, after I had exhausted current "standard" medical resources. I believe we should utilize every resource we have, including 20th century Western medicine, when it comes to getting our positive flow of energy back. In other words, when the Doctor repaired my eye with the laser years ago, I returned to a positive state of energy flow. This in turn allowed me the time to spiritually grow, and be open to alternative methods. We are constantly in transition. To throw out current medicine would not only be silly, but could be physically harmful. It is always best to use all available resources, as long as we can stay in that positive flow of energy, which is the foundation of all life.

Why do I have this affliction?

This was the most difficult stage to deal with. However it did lead me to a some answers. Once I remembered that the body gets all its life from one source –the connection to the Universal Energy flow. I concluded that I must have been cutting off my energy. I examined my current life and came up with some areas that needed attention. My Inner-Being had sent me a wake-up call, and I responded.

As our visions and paths change, *we have to keep up with them, or our energy may decrease.* Every path and vision is different and will require its unique amount of energy. The key here, is to be situationally aware of our visions and desired paths.

Do I want this affliction?

This appears to be a simple question, but is it. If we lack situational awareness, and are not on our life's path, a physical ailment will usually cause some directional change in ourselves, or in those around us. These are some questions I asked myself when facing blindness in one eye.

If I decide to keep this ailment, what would my new life be like?

Do I want to be forced into a new lifestyle?

Will I accept the pity, that friends often express out of kindness?

Will I enjoy playing catch with my children, without depth perception?

Is there anything I really want in this new lifestyle?

My answer was a very strong "no" to all of the above—and it felt right. Though I knew my ailment was something I had to go through, I also knew it would pass.

I do not want this affliction. I want to be healthy!

Now that I had made the choice to be free of the ailment, I began to *know and feel* that I would have my vision back. I had made the choice.

I had to seek a positive environment at all cost.

Once I knew my eye was going to return to normal, I wanted to be sure I did not expend any unnecessary energy defending my belief. I only informed certain people of my condition—people who would react positively. In other words, they would see my condition as an opportunity for growth, and not as an obstacle in my life.

In addition, others volunteered their energy. An entire order of Fransiscan Nuns added me to their prayers. Positive energy was flowing towards me from all directions. I asked everyone to see me—*only as well.*

A friend gave me water from Lourdes. I have known that certain areas of the planet have different energy levels, and had heard of the miracles in such places. I had also heard of many who were not cured in these locations. I concluded anything that enhances the power of belief inside a person, will have benefits.

I *used a new and specific affirmation, during meditations.*

Another friend gave me an affirmation that was special to them—a simple short poem, no more than three sentences. I used this daily during my morning meditation, instead of the one I normally use. This helped clear my mind in a different fashion, with more focus on my new situation.

Along with this meditation, I clearly envisioned two very strong, loving hands surrounding the black cloud in my eye. (These can be the hands of anyone you wish.) I saw the hands infuse the cloud with powerful energy, causing it to disappear. Then I envisioned the same hands replacing the area with new healthy cells.

I *visualized a healthy life ahead.*

Next, I began to visualize myself in a life with healthy vision. In other words, I saw myself doing those activities that I would normally do *with depth perception.* Here are some examples I used:

Watching an eagle land in a nest from a distance

Playing catch with my children

Flying airplanes

Bike riding

Jogging

Driving a car

As I did these visualizations, I *felt the emotion* in each activity. I did this several times a day.

I *appreciated all that was good in my life now.*

After I finished meditating and visualizing, I took a look at all the wonderful things in my current life. I sent out feelings of appreciation for them. I found that this helped keep me in that place of positive attraction. This feeling of appreciation was so strong, that I immediately recognized it again—right after my vision returned to normal!

18

In Search Of A New Paradigm

If I can attract better life experiences for myself, does that mean I can help others improve their lives?

We are each co-creating with the help of our powerful Inner-Being. I am not inclined to judge any person's experiences and decide whether or not they are attracting the "best" thing for them. All of our Inner-Beings are having life experiences on this physical plane. Only we can decide to enjoy the process.

However, we can help others attract different experiences. The very best way, of course, is to show them through your own life. They will now have something positive to focus on, and if they continue to do so, they will enhance their lives. More important-ly, they will want to continue focusing on the positive because of the many new benefits.

If you have attracted less fortunate people into your experi-ence, you will decide how you want to help them. You decide what brings you joy, and in what manner you may influence another person's environment. Don't be under the illusion that everyone who's in need of a positive example is living in the street—many are right down the hall in your office building.

When we do give to others, we must not pity them. If we do, they will invariably have a tougher time creating a better reality for themselves because energy is focused into their hardship.

Be aware that these very same people also attracted you into their experience. So if you are going to be a positive focal point for them, you can only do it by being positive. Help them focus on their possibility of a brighter future. Let them see *in you* that it can happen.

For this reason, I believe that when a society creates self-improvement programs the focus must be on the positive results desired. For example, we could call our national "Anti-Drug" campaign, the "Abundant Life" campaign.

What thoughts occur when you say "anti-drug?"

What thoughts occur when you say "abundant life?"

Now think about an entire society attracting these same thoughts. By focusing on the positive, we just might attract solutions more quickly to some of our long-standing social problems.

Will the real love please stand up.

Although love seems to be a very difficult concept to define, I believe it is a positive emotion from our Inner-Being, and should *feel good*. If it does not, as we mentioned earlier, we may be passing judgement, or attempted control unto another.

If you have experienced unconditional love, you know it always brings positive emotion. The reason is simple—when we unconditionally love, we allow others to be exactly who they are. We allow them to attract the experiences they want. We experience joy when we see them creating in the physical; knowing that, like ourselves, they are powerful spiritual entities. And we know all this because our Inner-Being is sending us that love feeling with positive emotion.

I am not talking about tolerating or putting up with someone.

When we tolerate another person's endeavors, we usually feel negative. Our Inner-Being sends us that negative emotion to warn us that judging others is not good for us, nor for them. Judging others cuts off *our* flow of positive energy from the Universe.

In short: When we love unconditionally, *we* are happy.

Can we control our current personal environment?

Our current "environment" is a unique arrangement of conditions created in this particular space-time. We attracted an experience that brought with it those very conditions. The Universe responded by delivering them to us. Positive thinking will not change *them*. But if we do not want a particular experience anymore, we can focus on new experience, and the conditions, or environment, that accompany that experience will be delivered to us.

How do I meditate?

There are many degrees of meditation, and even different definitions. But I will focus on one type which is very easy to do. You won't have to join a monastery or convent, and shaving your head will only get your scalp sunburned.

First and foremost, you will need about twenty minutes a day for yourself. Find a room where you will have absolutely no distractions. You need a comfortable chair —a recliner works nicely. Keep the lighting low, but adequate for moving about. You have just completed the most difficult part—getting rid of distractions!

Next, sit down and relax. State the purpose of your meditation. Do you want to do it just for relaxation? If so, then say that to yourself now. If you want thoughts to be attracted to you regarding a certain topic, state that now. Once you know your focus, your Inner-Being can help out tremendously.

Take three deep, slow breaths. When you're done, begin breathing normally again, except now count along with your breathing. In your mind, count three seconds while breathing in, and then one second holding it in, and two or three seconds out. For example, count 1-2-3 while breathing in, and count 4 while holding it in, then 5-6-7 while letting it out. These are not deep breaths; they are normal breaths that you have simply decided to monitor. If you breathe too deep or hard for any extended period of time, you could hyperventilate.

While you are calmly monitoring your breathing, focus on one small thing in life – it doesn't matter what. This is how we clear our mind. Let's face it, we are not exalted beings living on top of some mountain, trained since childhood to stop our mind from having thoughts. On the contrary, we've been taught to maximize our thinking and, for the most part, to keep our brains quite busy. So, when we put all of our attention into one small focal point, we clear the remainder of the mind. (Incidentally, this is why knitting, model building, and other similar hobbies are relaxing.)

Some good mind-cleansing techniques during meditation include:

Number counting.

Short phrases that coincide with your breathing,

"I am (breathing in), strong and healthy (breathing out)."

Humming "OHM" while breathing out.

You can choose whatever works best for you. Instead of counting, you may decide to simply focus on a quiet place that has special personal meaning. Do this for about ten to twenty minutes. If you try to meditate any longer, you will probably get frustrated, as our minds do not like being quieted for extended periods of time. You will receive more benefits by doing this exercise for fifteen minutes every day than you would by punishing yourself with two or three, one hour periods a week.

I believe you will enjoy results in just a few days. Not only will you be able to think with more clarity and focus, you will also have increased energy while maintaining a calmer demeanor.

How do I align, or "blend," with my Inner-Being?

Though the benefits are wonderful, the procedure is surprisingly quite simple. It can happen while you are meditating. As with any meditation, state the purpose before you begin. For example, you could say, "I want to blend with my Inner-Being." Next, meditate as you normally would and clear your mind—this is all you have to do. Your Inner-Being will do its work now by vibrationally aligning your energies with its own. You may feel some tingling or a few small itches here and there. Try to ignore them; they are simply the result of your physical body changing vibrational frequencies. Do this exercise every morning if you can. It is a dynamic process which will never end as you are in a constant state of realignment.

Note: You are only aligning your physical vibrational state with its corresponding spiritual vibrational state. You will not actually be vibrating at the same spiritual level. These are two different "states" which are the result of us living in space-time.

Again, the benefits are wonderful! You literally tap into your fundamental source of energy, and you feel it in so many ways. Sensitivity to your emotions, or guidance system, will increase, thereby empowering you with a stronger gut feeling that lets you know when something is beneficial for you. You will make better, more informed decisions! All this will happen because you are now facilitating the will of your Inner-Being—your broader, enlightened, and older self which is really you!

This kind of sounds like, "if it feels good, do it," so how could this ever be good for me, or even for an entire society?

The critical difference between the old paradigm and the new one is your ability to have situational awareness during the entire process. You must be able to recognize what you do not want. Your Inner-Being will guide you with emotion during your visualization procedure and attract other thoughts and images into your mind which are compatible with that vision. But you must monitor how you feel during the entire process.

Let's take a look at several examples of using your emotions as your most reliable guidance signal. Invariably, someone will always offer some form of the following scenario:

"Well, if I just do what feels good, I'd probably start off by not going into work today. Then I would tell my wife that this is my fun day, and I'm going to share it with my cute twenty-year-old secretary. The two of us may have a great lunch together at an expensive restaurant, but who cares? I just won't pay the bill when I get it, then...."

This is not responsible thinking! This is not what we are talking about. We are talking about taking responsibility for prepaving a life for yourself in the future which involves numerous intentionally attracted experiences. In other words, this fellow is attracting other circumstances and conditions with his vision, which he may not be aware of.

For example, is he prepared to lose his wife? What happens if the cute secretary drops him tomorrow? How will he feel when his wife doesn't take him back, or would she? Does he care either way? Will he miss his wife's great qualities? Are her "bad" ones really that bad? Does he want to live in the same house with his children?

These newly attracted experiences would be different for everyone, even with a similar scenario. In fact, one man may learn that he really is happy with his life and appreciate it more than before the visualization. Then again, someone else may come to a different conclusion. It would depend on what is really desired and the level of situational awareness.

I would submit that societies begin to deteriorate when its members are not allowed to focus on what they want. Whether held back by actual political barriers, or by their own sense of impotence, their feeling of powerlessness will affect that society.

Can I attract a specific person into my life?

We attract life experiences, and with those life experiences come associated conditions and environments. So, if we want to be in a relationship with a person, that is a life experience. The Universe will invariably present someone who fits your vision of a great relationship. Be careful, though, if you target someone first, then see the two of you in a relationship, you're doing it backwards. This often results in a relationship requiring "a lot of work". You cut the Universe off by not letting it match you up vibrationally. You may pick someone at the wrong frequency, which means he or she may not care to be in a relationship with you, or may simply not provide the joyous interaction you really desire.

Why do I keep attracting the same type of mate?

Many of us go through life attracting the same type of mate even though we could honestly say we want something different. Whether it is in a dating situation, or a second marriage, it all starts in the same place –our thoughts. We must ask this question: "Does my vibrational state match what I am seeking?"

Do I have to find my soul mate to have a healthy relationship?

Whether you have a relationship with your soul mate, or with a non-soul mate, is not the real question; it is how you grow from the experience that counts. People who have had unhappy relationships often think that if they find their soul mate, they will have that "great and cosmic" lifelong love affair. Although this is possible, this is not guaranteed. I believe this thought pattern developed because of our difficulty to accept responsibility for our own happiness. *The person we are now is the only one who can make us happy.* A soul mate could actually be in a relationship with us now, but because of our lack of situational awareness we are not benefiting from the experience. In other words, are we evoking the qualities we desire, or are we attacking the ones we dislike? After all, a relationship is the best place for us to learn who we are!

How does a jet fly?

An unrelated question? Maybe. The point here is that, just because we do not completely understand something, doesn't mean that it does not exist. Orville and Wilbur Wright knew that man could fly. Because of their vision the world has evolved into an entirely different reality. Do you think the two brothers were ridiculed when they were building that first airplane over ninety years ago, in a bicycle shop, no less! Of course they were! People will always be uncomfortable with vibrational states they do not understand.

Jets fly because air holds them up. Lift is created on an airfoil (wing) by its design. The top of the wing has a camber or curve which creates a low pressure area on top of the wing – and the whole plane is literally pulled into the sky! It is all very simple once you attract the appropriate thoughts. The question is not whether something exists or not, it's *when do we want to think the thoughts that put it into our reality.* The truth is, we could have started flying in any century.

If more people knew they could attract their own life experiences, what would happen in society?

Society is only as strong as its individual members. If we all knew how to attract our desired experiences, we would not only be stronger individuals, society as a whole would be uplifted.

Is it possible to visualize an experience that is too grand or big for me?

The Universe has no problem fulfilling any of our requests – as long as our vision is created with situational awareness, focus is maintained on the future desired experience, and we allow our environment to align – all from a positive point of attraction. This is why we must know our current limitations, i.e., what is required to keep us in a positive state of attraction.

How much of your current environment can you give up, and still remain in a positive state? One analogy might be, if you sell your car to buy a house, but are now unable to get to work in order to earn money, you won't be able to pay for the house.

Sin Gas?

While traveling through the mountains of Peru in search of "authentic" Shamans, I always made sure I carried an adequate supply of bottled "Agua Sin Gas," or water without carbonation. The word "sin" in Spanish literally means "without."

We all have the option of being in tune with the flow of Universal Energy, or being without that flow. Stated another way, we can choose to be in a position of sin (without), versus being in a position of harmony (with), the flow of Energy. As the captain of your life, you must do whatever it takes to get from a position of "sin," or lack of energy, to a position of connectedness.

Do not associate sin, however, with the guilt that society has built into it. Guilt is a negative emotion we get with a lack of Energy flow, signaling us that our thoughts are not focused on a positive co-created future. Guilt is not healthy. If we feel that we are in a state of sin, or low energy flow, we must recognize our negative thoughts, and replace them with positive ones. By doing so, we will open our energy valve and restore that positive point of attraction we desire.

Recognizing that the healing of guilt may be difficult, many religions have created rituals designed to focus our mind on cleansing. If we subscribe to these rituals, they can work for us.

For example, some South American Shamans (priests), utilize a combination of counter-clockwise body rotations, flint rock sparks, and chants, all done near a "mesa" or alter, under a midnight sky. The belief of those involved is intense and sincere, and I have no doubt it's as cleansing as the Catholic "confession" I was raised with.

The most important aspect of any healing is the release of negative thought patterns and the replacement by positive thought patterns. This is what accelerates energy flow and removes us from the place of sin, or without.

When we forgive another for what we perceive as sins against us, *we are healed*. Let's say a business partner wrongs you in your joint venture. If you dwell on the negativity of the situation, you will will feel bad because of the lack of energy flow. But if you forgive your partner, you will be able to replace the negative thoughts with ones of a positive, co-created future. You will then be back in tune with the energy flow and you will begin to feel good! All this results in more energy in your life and more personal growth. It certainly seems a more productive and happier way to live, than wallowing in pity or obsessing on vengeance.

Since you are in a position of strength after forgiving someone, you will be able to make more intelligent decisions regarding any future dealings. They may include refinement of your joint vision to avoid any more communication snags, or you may even find out that you need to dissolve the relationship. In either case, you will be in command of your experience, truly your own captain in tune with a powerful energy flow.

Why Am I On The Train?

A teacher of mine once told me, "It's not the destination that's important, but the train ride itself." I am certain he was advising me of the value of living in the moment—only my "moment" was preceding a "train ride" that I did not want to be on! Nor did I particularly want the destination of that train. Of course, I then felt guilty because I was one of those guys who thought he couldn't live in the moment.

Today I'm on a different train. I am now starting to live in the moment. And I must say, at those times when I truly experience the present, I feel great! I just became aware of living in the moment after I had developed a strong belief that the Universe would grant me co-created future experiences. I seemed to naturally rejoice in the present once I knew my future visions were secure.

When I look back, beginning with my college years, I realize that I always saw myself becoming an airplane pilot. And my vision held true. But deep inside, I now feel I missed out on something else, an aspect of life I once experienced as a child and am just now rediscovering: *living in the moment.*

As soon as I passed my first solo, I started working on my private pilot license, which led me to the cross-country phase check, followed by the private pilot checkride. I was then able to begin 150 hours of commercial instrument training. There would be several checkrides in this phase as well. Soon I would need a checkride on a "complex" single engine airplane, thus allowing me to enter the multi-engine training program.

In order to build the flight time and experience required for additional ratings, I had to be certified as a Flight Instructor. A year later, I applied two more ratings to my certificate with the completion of multi-engine instructor training and instrument instructor training. For three more years I anxiously built time through charter flying and flight instructing. After 1500 hours of total time, the government allowed me to take an Airline Transport Pilot test. Now I was ready to work for several commuter airlines. In 1983, this meant moving every six months, I lived in over a dozen cities.

And what do I remember, most of all? I remember waiting for my next achievement, my next rating, my next city—never really appreciating the moment. And I will never get those moments back. They existed only once for me in that space-time.

I am resolved never to let that happen to me again!

Armed with new, liberating information, I enabled myself to write this book. **Captain's Discretion** is my vision. I saw copies in bookstores before I even finished the first chapter. I allowed my environment to align so the vision would manifest. I focused energy and the Universe delivered.

Sure, there were many sixteen-hour days holed up in a cabin in the Georgia Mountains where I did nothing but write and meditate, but it was work I truly enjoyed, living each moment. That's the difference. I had never plunged into something so intensely, and yet with so much fervor and passion.

And right there is the difference between my former vision of becoming an airline pilot, and my current one of writing a book. This time, I have enough faith to live in the moment while I watch my future manifest. It is faith, finally, along with confidence in my future attracted experiences, that empowers me to live in the moment. And what a wonderful place to live. It is where everything in space-time is happening! *Trusting the Universe to continue with its current mode of operation, is the foundation of living in the moment.*

19

So What Do I Do Now?

Although pilots may follow the same set of procedures, they each individually utilize their own technique to accomplish them. An appreciation of each pilot's *uniqueness* is necessary in the cockpit, a concept which also holds true in life. What works for one person may, or may not work for another. It would be impossible to create a technique that would work for everyone, and there is good reason for that. The thoughts contained in each of us are so unique, their interaction and projected outcomes could never be calculated. There are infinite possibilities.

However, it does help to have a starting point. So in the interest of leaving you with a "launch pad," I will share some techniques I have found repeatedly successful.

Before we begin, let's update our situational awareness. Turn back to chapter five, and review your answers to the Destination Checklist. When you have finished, return to this page.

Thinking within your new paradigm, do you like what you read?

If you do, GREAT! If you do not, it is up to you to begin the exciting process of changing it.

Wake up earlier than normal tomorrow morning – preferably one hour, but not less than thirty minutes. Plan on being alone, with absolutely no distractions. From now on, this is your time. (It always was, but you were sleeping!) Find a comfortable chair to sit in – a recliner works nicely. You are going to focus on four separate areas during this time:

1. Your future long-term vision.
2. Your future short-term vision.
3. The day you are about to have.
4. Meditation.

Future Long-Term Vision

See yourself as you are in the future, visualizing in "broad strokes." For example: You are currently working in a large department store and you are not happy with the job. You think you'd like to be a school teacher – a math teacher, to be precise. So why wait? See yourself already there in the classroom, chalkboard full of numbers, students looking on, etc. If your Inner-Being sends you positive emotion, not only is it right for you, but it's on the way! This stage of the visualization powerfully expresses your personal mission statement, which you have already thought out. If you are presently developing your mission statement, however, or do not know what you want yet, the time should be used daily to monitor your emotions as you try on "new lives." Again, this is your time and, more importantly, your life. So be sure you see yourself happy!

Future Short-Term Vision

After ten or fifteen minutes of visualizing a year or two ahead, begin to focus your thoughts on the next month. See yourself having experiences which lead you into your future – meeting people who will help you, etc. See yourself in positive situations where things are starting to pop up to help you create your envisioned future. Be sure to see *yourself* in the script. Get to that "feeling place" where you are an actual part of the scenario. Know the Universe will deliver what you need at just the right time, perfectly matching up the right conditions with your desired experience.

The Day You Are About To Have

Take five or ten minutes to see yourself in the day you are about to attract. Visualize your experiences as happy ones: your drive to work will be safe and enjoyable; your professional and social interactions will be satisfying. Go through any meetings you may have that day, and see yourself at your best. (Thoughts will come to you all during the day, adding to your visualization.) Remember, it's your day. You are the captain.

Meditation

As I emphasized earlier, the purpose of the meditation must be known before you begin. I am particularly clear and energetic in the morning so I use this meditation time to align with my Inner-Being. In addition, when you've just finished focusing on your vision, your Inner-Being will now have a well defined target.

CONCLUSION

Community

We've discussed two types of balance in **Captain's Discretion**. The first is a *spiritual balance* which is the balance of energy flow and creation. Energy will always flow to where it's directed, then it will find expression. The resultant creation always balances with the amount of energy expression allowed. Nothing exists without this life-energy and there is always creation, where it is. *The Universe is always in balance.*

From our spiritual (Inner-Being) perspective, the world is also in balance. This is because our happiness comes from within, through our emotions, and is an attainable state for us to enjoy. It is a sign of strength as well, since our happiness is indicative of a higher energy flow into us. *From this perspective, we can make a difference in the world, because when our reality becomes a truly joyful one, we naturally want to share it with others.*

Physical balance is the perception from the physical plane that we need in order to fine-tune our energy flow. A feeling of contrast causes us to readjust our thinking, sharpen our focus, and co-create with the overall desire of our Inner-Being. In essence, we may be finding out what we do not want. As we do this, we can begin attracting those new life-enriching realities. But if we only have a physical perspective, we will never understand why we are attracting certain realities, nor how to get out of them, and we will live in a state of confusion. We won't see the big picture. The world will appear out of balance.

There is something else I would like to share with you. While witnessing the implementation of an Advanced Qualification Program at an airline, something amazing happened.

This particular airline did not have a Total Quality Management (TQM) program in place, which means there was little employee empowerment in the rank and file. Airlines have always been run in a quasi-military organizational manner, and creating TQM would be introducing something wildly different from the old chain of command. Adding to the barriers was airline competition so fierce that cost-cutting, layoffs, and even more top-down control seemed mandatory.

But as a "new way of doing business" was maturing in the flight operations department, TQM began to catch on. People enjoyed this more cooperative way of working together – being part of a team, instead of protecting turf, and empire-building. Soon people came from all departments and began to want the same thing. Team training spread throughout the company. The maintenance people, for example, had already begun their own version of TQM; a prairie fire had started that could not be stopped. The new way of business now was the *only* way of business.

When we become the captain of our life, we can make a difference in the world. And we can do it by simply being who we are: a unique expression living here on earth with a plan and a purpose.

We can be our own captains at home, in our community, at work, and in our organizations. Within these structures, we can physically manifest what is attracted by the society as a whole. The Universe is continually aligning our environment, based on our state of attraction. It truly is at our discretion to choose the desired state.